MATHS CONNECT 2 G

Sue Bright

Dave Kirkby

Lynne McClure

Catherine Roe

Bev Stanbridge

heinemann.co.uk
✓ Free online support
✓ Useful weblinks
✓ 24 hour online ordering

01865 888058

Heinemann

Inspiring generations

Heinemann Educational Publishers
Halley Court, Jordan Hill, Oxford OX2 8EJ
Part of Harcourt Education

Heinemann is a registered trademark of
Harcourt Education Limited

© Harcourt Education Ltd

First published 2004

07 06 05 04
10 9 8 7 6 5 4 3 2 1

British Library Cataloguing in Publication Data is available
from the British Library on request.

ISBN 0 435 53493 9

Designed by Bridge Creative
Typeset by Tech-Set Ltd, Gateshead, Tyne and Wear
Original illustrations © Harcourt Education Limited, 2004
Illustrated by Tech-Set Ltd and Bigtop Design
Cover design by mccdesign ltd
Printed in Italy by Printer Trento srl

Acknowledgements
Every effort has been made to contact copyright holders of material reproduced in this book. Any omissions
will be rectified in subsequent printings if notice is given to the publishers.

Consultant
Jackie Fairchild

The authors and publishers would like to thank the following for permission to use photographs:
p13 Getty Images UK/Photo disc, p151 Getty Images UK/Stone, p191 Corbis/Yann Arthus-Bertrand

Cover photo: Getty images ©

Publishing team

Editorial		Design	Production
Amanda Halden		Phil Leafe	Siobhan Snowden
Naomi Anderson			
Lindsey Besley	Ian Crane	**Picture research**	
Lauren Bourque	Carol Harris	Jane Hance	
Bryony Costin	Katherine Pate	Bea Ray	
Melissa Okusanya	Margaret Shepherd		
Maggie Rumble	Laurice Suess		
Nick Sample	Chris Worth		

Tel: 01865 888058 email: info.he@heinemann.co.uk

MATHS CONNECT 2 G

Contents

Contents   iv

How to use this book   viii

Unit 1 N/A1	Integers and sequences	2
Unit 2 SSM1	Angles and shapes	14
Unit 3 HD1	Probability	26
Unit 4 N2	Fractions, decimals and percentages	38
Unit 5 A2	Equations and formulae	50
Unit 6 SSM2	Measures, area and perimeter	62
Unit 7 A3	Functions and graphs	74
Unit 8 N3	Numbers and calculations	86
Unit 9 SSM3	Transformations	104
Unit 10 A4	Solving equations and using formulae	116
Unit 11 HD2	Tables and statistics	128
Unit 12 N4	Squares, brackets and calculations	140
Unit 13 A5	Sequences and graphs	152
Unit 14 N5	Ratio and proportion and solving problems	168
Unit 15 SSM4	Exploring 2-D and 3-D shapes	180
Unit 16 HD3	Applying skills and analysing data	198

Index   212

Contents

N/A1 Integers and sequences

1.1	Adding integers	2
1.2	Subtracting integers	4
1.3	Tests for divisibility	6
1.4	Sequences from patterns	8
1.5	Generating sequences	10
1.6	Investigating sequences	12

SSM1 Angles and shapes

2.1	Angle sums	14
2.2	Names of angles	16
2.3	Measuring and drawing angles	18
2.4	Drawing triangles 1	20
2.5	Parallel and perpendicular lines	22
2.6	Angle calculations	24

HD1 Probability

3.1	Probability using words	26
3.2	Probability using numbers	28
3.3	Possible outcomes	30
3.4	Tallies and frequency tables	32
3.5	Estimating probability	34
3.6	Comparing probability	36

N2 Fractions, decimals and percentages

4.1	Equivalent fractions	38
4.2	Fractions and decimals	40
4.3	Ordering fractions	42
4.4	Adding and subtracting fractions	44
4.5	Fractions of amounts	46
4.6	Percentages	48

A2 Equations and formulae

5.1	Simplifying algebraic expressions	50
5.2	Expanding brackets	52
5.3	Substitution	54
5.4	Solving equations	56
5.5	More solving equations	58
5.6	Formulae	60

SSM2 Measures, area and perimeter

6.1	Multiplying and dividing by 10, 100, 1000	62
6.2	Scales and measures	64
6.3	Area of rectangles	66
6.4	Perimeter and area	68
6.5	Area of triangles	70
6.6	Surface area	72

A3 Functions and graphs

7.1	Mappings	74
7.2	Identifying mappings	76
7.3	Special graphs	78
7.4	Graphs	80
7.5	The y-intercept	82
7.6	Gradients	84

N3 Numbers and calculations

8.1	Converting between metric units	86
8.2	Ordering decimals	88
8.3	Rounding	90
8.4	Adding and subtracting	92
8.5	Multiplying	94
8.6	More multiplying	96
8.7	Multiples	98
8.8	Factors	100
8.9	Prime numbers	102

SSM3 Transformations

9.1	Reflection	104
9.2	Rotation	106
9.3	Reflection and rotation symmetry	108
9.4	Translation	110
9.5	Repeated transformations	112
9.6	Drawing enlargements	114

A4 Solving equations and using formulae

10.1	Algebraic expressions	116
10.2	Why simplify?	118
10.3	Solving equations involving divisors	120
10.4	Using equations	122
10.5	Formulae in words	124
10.6	Constructing formulae	126

HD2 Diagrams and statistics

11.1	Two-way tables	128
11.2	Averages	130
11.3	Frequency tables and calculating the mean	132
11.4	Comparing two distributions	134
11.5	Grouping data	136
11.6	Calculating statistics	138

N4 Squares, brackets and calculations

12.1	Squares and square roots	140
12.2	Brackets	142
12.3	Division	144
12.4	Decimal division	146
12.5	Mental methods 1	148
12.6	Mental methods 2	150

A5 Sequences and graphs

13.1	The general term	152
13.2	Exploring sequence patterns	154
13.3	Finding the general term	156
13.4	Spreadsheets	158
13.5	Conversion graphs	160
13.6	Drawing graphs	162
13.7	Interpreting graphs	164
13.8	Graphs in real life	166

N5 Ratio and proportion and solving problems

14.1	Proportion	168
14.2	Ratio and proportion	170
14.3	Ratio	172
14.4	Solve it!	174
14.5	Multi-step problems	176
14.6	Which strategy?	178

SSM4 Exploring 2-D and 3-D shapes

15.1	Drawing triangles 2	180
15.2	Constructing triangles	182
15.3	Shapes on coordinate grids	184
15.4	Mid-points	186
15.5	Shapes and paths	188
15.6	Visualising 3-D shapes	190
15.7	Constructing nets 1	192
15.8	Constructing nets 2	194
15.9	Volume	196

HD3 Applying skills and analysing data

16.1	Planning an investigation	198
16.2	Collecting data	200
16.3	Displaying data	202
16.4	Line graphs	204
16.5	More about pie charts	206
16.6	Drawing and interpreting diagrams	208
16.7	Communicating results	210

Matching charts linking the content of the lessons in each Unit to the *Sample medium-term plans for mathematics* and the Year 8 teaching programme from the *Framework for teaching mathematics* are available on the web at **www.heinemann.co.uk**

How to use this book

This book is divided up into 16 colour-coded Units. **Algebra** Units are green, **Number** Units are orange, **Space, Shape and Measure** Units are blue and **Handling Data** Units are red. Each Unit is divided up into lessons. Each lesson is on its own double page spread.

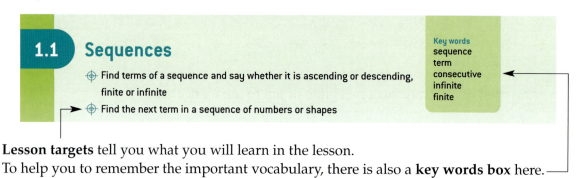

Lesson targets tell you what you will learn in the lesson.
To help you to remember the important vocabulary, there is also a **key words box** here.

A number **sequence** is a set of numbers in a given order, e.g. 1, 2, 3, 4, 5, …
Each number in a sequence is called a **term** .
Terms next to each other are called **consecutive terms** .
Sequences may be **ascending** (e.g. 2, 4, 6, 8, …) or **descending** (e.g. 18, 15, 12, 9, …).
The sequence 1, 2, 3, 4, 5 … is **infinite** . We could go on counting forever.
The sequence 10, 12, 14 … 98 is **finite** . The dots mean that there are missing terms and that the sequence continues in the same way until the final value, 98, is reached.

In the **explanation box**, you can see a summary of the key ideas that are covered in the lesson. The key words are highlighted in yellow.

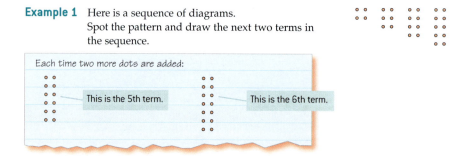

The **worked examples** show you methods of answering the exercise questions. On the blue paper, you can see the kind of working you should be writing in your exercise book. The **hint boxes** help to explain how you can calculate the answers.

The **exercise** for each lesson is made up of three types of question:

- **practice** questions, which allow you to practise the basic skills
- **problem** questions, which encourage you to apply the skills you have learned
- **investigation** questions, which give you practice at solving open-ended problems.

The following features are found in the exercise:

Consecutive terms are terms that are next to each other.

Hint boxes give tips and reminders to help you with the questions

Questions you should try without the help of a calculator are marked with this symbol

Questions that require you to use a calculator are marked with this symbol.

If there is no symbol, you can choose whether or not to use a calculator.

Adding integers

⊕ Know how to add positive and negative numbers

⊕ Know how to use the sign change key on a calculator

The set of **positive** and **negative** whole numbers, including zero, are called **integers**.
To add two integers, use a **number line** to help you.
Start on the line at the first number then:
- to add a positive number count right
- to add a negative number count left.

For example, to calculate $^+3 + {}^-5$, start at $^+3$, and count 5 jumps to the left.

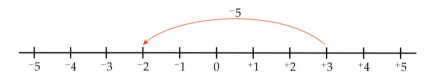

So, $^+3 + {}^-5 = {}^-2$

Example 1   Copy and complete these additions:

   a) $^+3 + {}^-6 =$ b) $^-5 + {}^+7 =$ c) $^-4 + {}^-2 =$

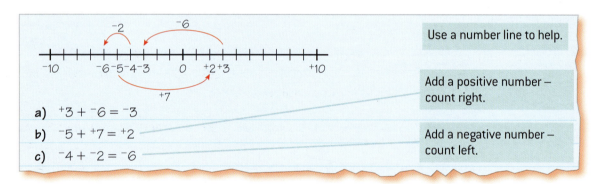

Use a number line to help.

a) $^+3 + {}^-6 = {}^-3$

b) $^-5 + {}^+7 = {}^+2$

Add a positive number – count right.

c) $^-4 + {}^-2 = {}^-6$

Add a negative number – count left.

Example 2   Use the sign change key of a calculator to find the value of:

   $^-24 + {}^-31 =$

$^-24 + {}^-31 = {}^-55$

Key in

Exercise 1.1

❶ Complete these additions.

   a) $^+3 + {}^-2 =$ b) $^-5 + {}^+4 =$ c) $^-2 + {}^+5 =$ d) $^-3 + {}^-2 =$

   e) $^-5 + {}^+5 =$ f) $^-7 + {}^-7 =$ g) $^-7 + {}^+13 + {}^-8 =$ h) $^+6 + {}^-12 + {}^+3 =$

2 In an addition pyramid, the number in each brick is found by adding the two directly below it. Copy and complete these pyramids.

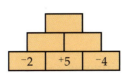

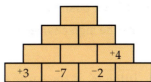

 3 Use the sign change key of a calculator to find the value of:

a) $^+37 + {}^-56 =$　　　　b) $^-98 + {}^+36 =$　　　　c) $^-72 + {}^-43 =$

d) $^+127 + {}^-246 =$　　　e) $^-243 + {}^+624 =$　　　f) $^-136 + {}^-242 =$

4 Copy and complete these addition tables:

+	$^-2$	$^+4$	$^-5$
$^-3$			
$^+7$			
$^-9$			

+	$^+3$	$^-4$	$^+7$
$^+2$			
$^-4$			
$^-8$			

5 Write the missing number in these calculations.

a) $^+4 + \boxed{} = {}^+5$

b) $^+3 + \boxed{} = {}^+1$

c) $^-5 + \boxed{} = {}^-2$

d) $^-6 + \boxed{} = {}^+3$

e) $^-2 + \boxed{} = {}^-9$

f) $^+4 + \boxed{} = {}^-3$

g) $\boxed{} + {}^+8 = 0$

h) $\boxed{} + {}^-7 = 0$

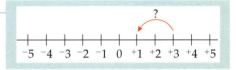

6 At 6 am the temperature is $^-3°C$ but by midday it has risen by 8°. What is the temperature at midday? By midnight the temperature has fallen by 15° from what it was at midday. What is the temperature at midnight?

7 Find the value of each letter, then convert the coded message to letters, and read the message.

$^+6 + {}^-4 = A$　　　　$^+9 + {}^-10 = B$

$^-4 + {}^-4 = D$　　　　$^+2 + {}^-6 = E$

$^-1 + {}^+4 = F$　　　　$^+10 + {}^-15 = G$

$^+13 + {}^-12 = H$　　　$^-1 + {}^+5 = I$

$^-10 + {}^+10 = L$　　　$^+5 + {}^-7 = M$

$^-8 + {}^+13 = N$　　　$^-3 + {}^-4 = O$

$^+2 + {}^-5 = P$　　　　$^+3 + {}^+4 = R$

$^-6 + 0 = S$　　　　　$^-8 + {}^-2 = T$

$^-6 + {}^+14 = U$　　　$^-10 + {}^+16 = V$

$^-5$	$^-7$			
$^+7$	$^+4$	$^-5$	$^+1$	$^-10$
$^-10$	$^-7$			
$^+2$	$^-8$	$^-8$		

$^-3$ $^-7$ $^-6$ $^+4$ $^-10$ $^+4$ $^+6$ $^-4$

$^+5$ $^+8$ $^-2$ $^-1$ $^-4$ $^+7$ $^-6$

$^-5$ $^-7$

0 $^-4$ $^+3$ $^-10$

$^-10$ $^-7$

$^+2$ $^-8$ $^-8$

$^+5$ $^-4$ $^-5$ $^+2$ $^-10$ $^+4$ $^+6$ $^-4$ $^-6$

Investigation

8 Use the digits 0, 1, 2, 3, 4 and 5 each time and positive and negative signs. How many different additions can you make with an answer of $^-2$?

1.2 Subtracting integers

◈ Know how to subtract positive and negative numbers
◈ Know how to use the sign change key on a calculator

Key words
integer
inverse
number line

Subtracting an **integer** is the same as adding its **inverse** .

The **inverse** of $^-3$ is $^+3$, so:

● subtracting $^-3$ is the same as adding $^+3$ ——— To add a positive number count right.
● subtracting $^+3$ is the same as adding $^-3$ ——— To add a negative number count left.

For example, to calculate $^+3 - {}^-2$, change it to $^+3 + {}^+2$, and then:
start at $^+3$ on a **number line** , and count two jumps to the right.

So, $^+3 - {}^-2 = {}^+3 + {}^+2 = {}^+5$

Example 1 Copy and complete these subtractions:

a) $^-3 - {}^+5 =$ b) $^+4 - {}^-2 =$

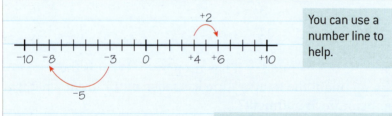

You can use a number line to help.

a) $^-3 - {}^+5 = {}^-3 + {}^-5 = {}^-8$ Add the inverse of $^+5$ (that is, $^-5$)

b) $^+4 - {}^-2 = {}^+4 + {}^+2 = {}^+6$ Add the inverse of $^-2$ (that is, $^+2$)

Example 2 Use the sign change key of a calculator to find the value of:
$^-321 - {}^-18 =$

$^-321 - {}^-18 = {}^-303$

Key in `3` `2` `1` `+/−` `−` `1` `8` `+/−` `=`

Exercise 1.2

1 Complete these subtractions.

a) $^+3 - {}^+6 =$ b) $^-5 - {}^+4 =$ c) $^+8 - {}^-2 =$
d) $^-6 - {}^-3 =$ e) $^+4 - {}^+5 =$ f) $^+3 - {}^-4 =$
g) $^-3 - {}^-3 =$ h) $^+7 - {}^+7 =$ i) $^-4 - {}^+3 - {}^+2 =$
j) $^+7 - {}^+3 - {}^-5 =$

2 In a subtraction pyramid, the number in each brick is found by subtracting the two directly below it, i.e. left number take away right. Copy and complete these pyramids.

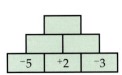
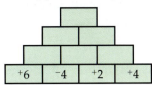

3 Use the sign change key of a calculator to find the value of:
a) $^+42 - {}^+36 =$
b) $^+78 - {}^-35 =$
c) $^-362 - {}^+124 =$
d) $^-834 - {}^-279 =$
e) $^+285 - {}^+384 =$
f) $^-621 - {}^+144 =$

4 Copy and complete these subtraction tables by starting with the number on the left and subtracting each number along the top in turn.

−	⁻3	⁺5	⁻6
⁺7			
⁻2			
⁻4			

−	⁺4	⁻2	⁺7
⁻3			
⁺5			
⁻8			

5 Find the value of each letter, then convert the coded message to letters, and read the message.

$^-2 - {}^-5 = A$ $^+6 - {}^+8 = B$
$^+4 - {}^-4 = C$ $^-6 - {}^-6 = D$
$^+3 - {}^+4 = E$ $^-2 - {}^+5 = H$
$^-7 - {}^-11 = I$ $^+6 - {}^+4 = N$
$^+9 - {}^+4 = O$ $^-8 - {}^-5 = R$
$^-4 - {}^-5 = S$ $^+5 - {}^-1 = T$
$^+15 - {}^+19 = U$ $^-8 - {}^-2 = V$
$^-4 - {}^+1 = Y$

⁺6	⁺5							
⁺1	⁻4	⁻2	⁺6	⁻3	⁺3	⁺8	⁺6	
⁻5	⁺5	⁻4						
⁺3	0	0						
⁺6	⁻7	⁻1						
⁺4	⁺2	⁻6	⁻1	⁻3	⁺1	⁻1		

6 A game for two players.
Use two cubes, one red and one blue. Write $^-2$, $^-1$, 0, $^+1$, $^+2$, $^+3$ on the red cube and $^-3$, $^-2$, $^-1$, 0, $^+1$, $^+2$ on the blue cube.
Take turns to roll the dice and then subtract the number on the red dice from the number on the blue dice. The player with the highest answer wins the round, and scores 10 points. Repeat for 10 throws each. The winner is the player with the highest total score.

7 Use the digits 0, 1, 2, 3, 4 and 5 each time and positive and negative signs. Make up calculations with answers from $^-15$ to $^+15$.
For example $^-15 = {}^-5 + {}^-4 + {}^-3 + {}^-2 + {}^-1$

Tests for divisibility

⊕ Know tests for divisibility for the numbers 2 to 10

One number is **divisible** by another number if it divides exactly by that number with no remainder.

The tests for **divisibility** are:

by 2 – is the last digit even?
by 3 – is the digit total a **multiple** of 3?
by 4 – do the last two digits make a multiple of 4?
by 5 – is the last digit 0 or 5?
by 6 – does it pass the test for 2 and for 3?
by 7 – can you partition it into known multiples of 7?
by 8 – halve it then test for divisibility by 4.
by 9 – is the digit total a multiple of 9?
by 10 – is the last digit 0?

To test for divisibility by a larger number, apply the tests for a pair of its factors. The pair should not themselves have common factors.

So to test for divisibility by 24, 3 and 8 are suitable but 4 and 6 are not. For example:

by 15 – apply tests for 3 and for 5. 3×5 is a **factor pair** of 15
by 18 – apply tests for 2 and for 9.
by 24 – apply tests for 3 and for 8.

Example 1   72   135   161   600

Which of these numbers are:
a) divisible by 8   b) divisible by 9
c) divisible by 6   d) divisible by 7?

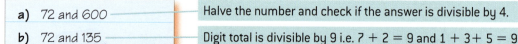

a) 72 and 600 — Halve the number and check if the answer is divisible by 4.

b) 72 and 135 — Digit total is divisible by 9 i.e. $7 + 2 = 9$ and $1 + 3 + 5 = 9$

c) 72 and 600 — The numbers are divisible by both 2 and 3.

d) 161 — We can partition 161 into 140 (divisible by 7) and 21 (divisible by 7)

Example 2   Is 315 divisible by 15?

15 has the factors 5 and 3
315 is divisible by 5 — The number 315 ends in 5.
315 is divisible by 3 — $3 + 1 + 5 = 9$ which is divisible by 3.
315 is divisible by 15

Exercise 1.3

1 78   108   147   200   405

Which of these numbers are:
- **a)** divisible by 2
- **b)** divisible by 5
- **c)** divisible by 3
- **d)** divisible by 6
- **e)** divisible by 4
- **f)** divisible by 8
- **g)** divisible by 9
- **h)** divisible by 7?

2 Copy and complete a divisibility test table for these numbers. 108 has been done for you.

35   38   45   189   480

Divisible by	Numbers	Test
2	38, 108, 480	Last digit is even
3	108	
4	108	
5		
6	108	
7		
8		
9	108	
10		Ends in 0

3 Use your table from **Q2** to find out which numbers are divisible by both:
- **a)** 5 and 9
- **b)** 5 and 7
- **c)** 3 and 7
- **d)** 2 and 5
- **e)** 2 and 9
- **f)** 2, 3, 4, and 5

4 Which numbers from **Q2** are divisible by:
- **a)** 35
- **b)** 10
- **c)** 18
- **d)** 21
- **e)** 45
- **f)** 120

5 **a)** Draw a divisibility test table for these numbers:

    42   45   70   120

    b) Use your table to find any numbers divisible by   **i)** 14   **ii)** 15

6 **a)** 165 is divisible by 15. Give one other number it is also divisible by.
    b) 126 is divisible by 21. Give one other number it is also divisible by.
    c) 96 is divisible by 16. Give one other number it is also divisible by.

7 **a)** Is 126 divisible by 18? ——— Check for divisibility by 2 and 9.
    b) Is 140 divisible by 15?
    c) Is 93 divisible by 12?

8 70 people go to a dinner party.
Each table seats 14 people.
Does everyone sit at a full table?

Sequences from patterns

◈ Generate a sequence from a pattern

◈ Explain how a pattern sequence grows

A **sequence** is a set of numbers in a given order. This is a sequence of cross **patterns** .

Term number	1	2	3	4	…
Number of squares	5	9	13	17	…

Generating a sequence means writing down the terms of a sequence.

The numbers of squares in the crosses generates a sequence: 5, 9, 13, 17, …

Looking at the pattern can help you see how the sequence grows.

Example   a) Draw the fifth term in the cross pattern sequence.

b) How many squares are there in the fifth cross pattern?

c) Explain how the sequence grows.

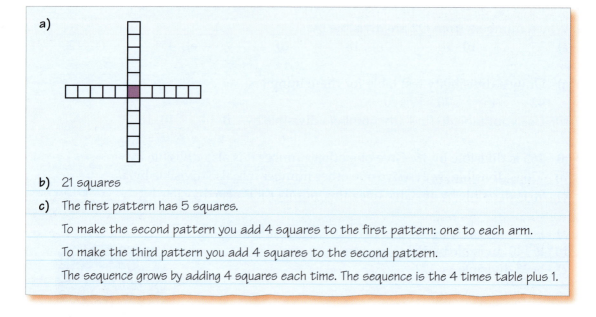

a)

b) 21 squares

c) The first pattern has 5 squares.

To make the second pattern you add 4 squares to the first pattern: one to each arm.

To make the third pattern you add 4 squares to the second pattern.

The sequence grows by adding 4 squares each time. The sequence is the 4 times table plus 1.

Exercise 1.4

❶ Draw the next two patterns in this sequence.
Explain how a pattern sequence grows.

2 Draw the next two terms in each matchstick pattern sequence.
For each sequence, explain how the pattern grows.

a)

b)

c)

3 For each pattern sequence:
  i) Explain how the pattern grows.
  ii) Write down the first five terms of the number sequence for the pattern.

a)

b)

c)

4 **Square numbers**
The first four patterns in the sequence of
square numbers are shown:

Term number	Number sequence
1	$1 \times 1 = 1$
2	$2 \times 2 = 4$
3	$3 \times 3 = 9$
4	$4 \times 4 = 16$

Look at the pattern:
1st term has $1 \times 1 = 1$ dots
2nd term has $2 \times 2 = 4$ dots

Write down the first ten terms of the number sequence.

Continue the pattern.

5 **Triangular numbers**
Here are the first four patterns in the sequence
of triangular numbers:
Draw the next two patterns in the sequence.
Write the number sequence for the pattern.

Investigation

6 You can draw square numbers as dots arranged in squares.   See **Q4**.

You can draw triangular numbers as dots arranged in triangles.   See **Q5**.

Show how you could draw the numbers in this sequence as dots arranged in rectangles:
3, 6, 9, 12
Find some other sequences of dots arranged in rectangle patterns.
Write down the number sequence for each pattern sequence.

Generating sequences

- ⊕ Generate a sequence given a starting point and a rule to go from term to term
- ⊕ Use the rule to find a term in a sequence without finding all the values in between

Key words
sequence
ascending
descending
consecutive
term
generate
term-to-term
rule

A **sequence** is a set of numbers in a given order.

A sequence can be **ascending** (going up) or **descending** (going down).

Consecutive terms are terms that are next to each other.

Generating a sequence means writing down the terms of the sequence. To do this you need to know the pattern that the sequence follows.

To generate a sequence you may be given a starting point and a **rule** that connects one term to the next. This is called the **term-to-term rule** .

For example, for the sequence 2, 4, 6, 8 the starting point is 2 and the term-to-term rule is 'add 2'.

Example 1   A sequence starts with 3 and the term-to-term rule is 'add 4'.
Find the first five terms of the sequence:

3, 7, 11, 15, 19, ...

$3 + 4 = 7$
$7 + 4 = 11$
$11 + 4 = 15$ ... and so on.

Example 2   The first term of a sequence is 2. The term-to-term rule is 'multiply by 3'.
What are the first three terms of the sequence?

2, 6, 18

$2 \times 3 = 6$
$6 \times 3 = 18$

Exercise 1.5

1 The first term of a sequence is 5. The term-to-term rule is 'add 3'.
Write down the first five terms.

2 A sequence starts with 100. The term-to-term rule is 'subtract 10'.
Write down the first four terms.

3 The first term of a sequence is 3. The term-to-term rule is 'multiply by 3'.
Write down the first five terms of the sequence.

4. Here are some starting points and term-to-term rules for sequences. Write down the first five terms of each sequence.

	Starting point	Term-to-term rule
a)	6	Add 4
b)	4	Multiply by 3
c)	50	Subtract 5
d)	3	Add 0.5
e)	1	Add 1 then multiply by 2

5. On his first day a man working in a Chocolate factory produces 120 chocolates.

   On the second day he produces 140 chocolates and on the third day he produces 160 chocolates.
   a) Write down how many chocolates he will produce on the fourth day.
   b) How many chocolates will he produce on the fifth day?
   c) Describe the sequence giving the first term and the term-to-term rule.
   d) Why can't the sequence continue in this way?

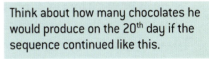

Think about how many chocolates he would produce on the 20ᵗʰ day if the sequence continued like this.

6. Dilip starts running to get fit.
   In the first week he runs 1 km.
   In the second week he runs 3 km.
   In the third week he runs 5 km.
   a) Write down the sequence of distances Dilip runs.
   b) If this pattern continues, how many km will he run in the fourth week?
   c) How many km will he run in the sixth week?
   d) Why can't the sequence continue in this way?

Think about how far he would run in the 20ᵗʰ week if the sequence continued like this.

7. The first term of a sequence is 5. The term-to-term rule is 'add 3'.
   a) Copy and complete the table below showing the sequence:

Term number	1	2	3	4	5
	2 + 3	2 + 3 + 3	2 + 3 + 3 + 3	……	……
	5	8	……	……	……

   b) How many times did you add 3 to the first term to find the:
      i) second term **ii)** third term **iii)** fourth term **iv)** fifth term?
   c) How many times would you need to add 3 to the first term to find the sixth term?
   d) What is the sixth term?
   e) How many times would you need to add 3 to the first term to find the tenth term?
   f) What is the tenth term?

Investigation

8. A sequence starts 6, 11, 16, 21, …
   a) Write down the next four terms in the sequence.
   b) What is the tenth term?
   c) How many times did you have to add 5 to get from the first term to the tenth term?
   d) How many times did you have to add 5 to get from the first term to the fifth term?
   e) Follow this pattern to find the twentieth term, without finding all the terms in between.

Investigating sequences

◈ Find any term in a sequence without finding all the terms in between

Key words
sequence
differences
consecutive

We can find the next few terms of a **sequence** in one of the following ways:
- by using the term-to-term rule.
- by looking at the **differences** between **consecutive** terms and spotting the pattern.

Look at this sequence: 5, 9, 13, 17, 21 …

The sequence is going up in 4's. Compare the sequence with the 4 × table plus 1.

Sequence	4 + 1	4 + 4 + 1	4 + 4 + 4 + 1	4 + 4 + 4 + 4 + 1	4 + 4 + 4 + 4 + 4 + 1

1st term is   **1** lot of 4, plus 1.
2nd term is   **2** lots of 4, plus 1.
3rd term is   **3** lots of 4, plus 1.
By following the pattern it is easy to find any term in the sequence.
For example, the **50**th term is **50** lots of 4 plus 1 = 201

Example 1   Find the 20th term of this sequence: 6, 11, 16, 21, 26 …

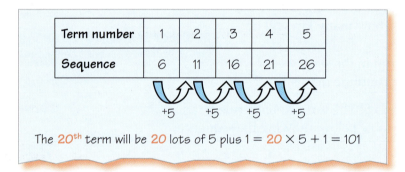

Term number	1	2	3	4	5
Sequence	6	11	16	21	26

+5   +5   +5   +5

The **20**th term will be **20** lots of 5 plus 1 = **20** × 5 + 1 = 101

The sequence goes up in 5's so we compare the sequence with the 5 × table plus 1. The **1**st term is **1** lot of 5 plus 1 so the **20**th term is **20** lots of 5 plus 1.

Example 2   The first term of a sequence is 48. The term-to-term rule is 'subtract 2'.
     a)   What is the 10th term?
     b)   What is the 20th term?

Term number	1	2	3	4	5
Sequence	50 − (1 × 2) = 48	50 − (2 × 2) = 46	50 − (3 × 2) = 44	50 − (4 × 2) = 42	50 − (5 × 2) = 40

To find the next term in the sequence we must subtract an extra 2 each time. To find the **10**th term, subtract **10** lots of 2 from 50.

a)   50 − (10 × 2) = 30

b)   50 − (20 × 2) = 10   To find the **20**th term, subtract **20** lots of 2 from 50.

Exercise 1.6

1 The first term of a sequence is 10 and the term-to-term rule is 'add 2'.

   a) Copy and complete the table below for the first five terms of the sequence:

Term number	1	2	3	4	5
Sequence	$8 + (1 \times 2)$ = ……	$8 + (2 \times 2)$ = ……	$8 + (3 \times 2)$ = ……		

   b) How many lots of 2 would you add to 8 find the 7th term?

   c) What is the 7th term?   **d)** What is the 10th term?   **e)** What is the 21st term?

2 The first term of a sequence is 17 and the term-to-term rule is 'subtract 3'.

   a) Copy and complete the table below for the first five terms of the sequence:

Term number	1	2	3	4	5
Sequence	$20 - (1 \times 3)$ = ……	$20 - (2 \times …)$ = ……	$20 - (… \times 3)$ = ……		

   b) How many lots of 3 would you subtract from 20 to find the 7th term?

   c) What is the 7th term?   **d)** What is the 10th term?   **e)** What is the 21st term?

3 The first term of a sequence is 13 and the term-to-term rule is 'add 7'.

   a) Write down the first ten terms of the sequence.

   b) Compare the sequence with the 7 times table.
      How many lots of 7 would you add to 6 to find the 19th term?

   c) What is the 19th term?   **d)** What is the 21st term?   **e)** What is the 30th term?

4 The first term of a sequence is 210 and the term-to-term rule is 'subtract 10'.

   a) Write down the first ten terms of the sequence.

   b) Compare the sequence with the 10 times table.
      How many lots of 10 would you subtract from 210 to find the 19th term?

   c) What is the 19th term?   **d)** What is the 21st term?   **e)** What is the 30th term?

5 For each of the following sequences:

   a) Find the next two terms   **b)** Find the twentieth term.

      i) 7, 9, 11, 13, …   **ii)** 5, 8, 11, 14, 17 …

     iii) 101, 99, 97, 95, 93, …   **iv)** 101, 91, 81, 71 …

> For each sequence, find the first term and the term-to-term rule.

6 The number of wild poppies in a field decreases each year.
An environmentalist records the number of poppies in the field each year for five years.

Year	Number of poppies
1	5000
2	4750
3	4500
4	4250
5	4000

   a) Look at the sequence of 'Number of poppies'. Describe the sequence by giving a first term and a term-to-term rule.

Assume the number of poppies continues to decrease at the same rate.

   b) Work out how many poppies will be left in the field after 10 years.

   c) After how many years will there be no poppies in the field?

Angle sums

⊕ **Know the sum of angles at a point, on a straight line and in a triangle**

Key words
angle
degree
angle sum

An **angle** is a measure of turn. Angles are usually measured in **degrees**, or ° for short.

Angles that meet at a point make a full turn so they add up to **360°**.

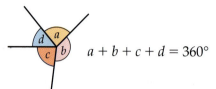

$a + b + c + d = 360°$

Angles that meet at a point on a straight line make a half turn so they add up to **180°**.

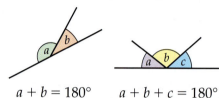

$a + b = 180°$ $a + b + c = 180°$

The three angles of any triangle add up to **180°**.

The **angle sum** of a triangle is **180°**.

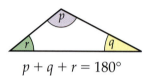

$p + q + r = 180°$

Example   Calculate (do not measure) the size of the angles marked with the letters a, b and c. Give reasons for your answers.

a) 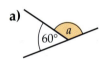   **b)**   **c)**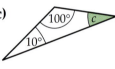

a) $180° - 60° = 120°$ so a is 120°.

Angles on a straight line add up to 180°.

b) $360° - 110° - 190°$

$= 360° - 300°$

$= 60°$ so b is 60°.

Angles at a point add up to 360°.

c) $180° - 100° - 10°$

$= 180° - 110°$

$= 70°$ so c is 70°.

The three angles of a triangle add up to 180°.

Exercise 2.1

1 Calculate (do not measure) the size of the lettered angles. Give reasons for your answers.

a) 70° a

b) b 115°

c)
50°
c

2 Calculate (do not measure) the size of the lettered angles. Give reasons for your answers.

a)

b)

c)

3 Calculate (do not measure) the size of the lettered angles. Give reasons for your answers.

a)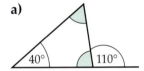

b)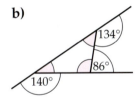

c)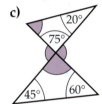

4 Use this diagram of a compass to calculate the following turns, all in a **clockwise** direction:

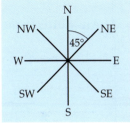

a) N to SE
b) N to SW
c) E to NW
d) N to NW
e) SE to NW
f) NW to SE
g) S to NE.

Write down two turns that are:

h) 90° clockwise
i) 45° clockwise
j) 180° clockwise
k) 135° clockwise.

5 Sketch these diagrams and fill in the sizes of all the marked angles.

a)

b)

c)

6 Ben is making a wooden puzzle.

The diagram shows four of the wooden pieces that fit together along one of the edges.

What is the size of the red angle?

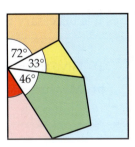

7 Use a ruler and a pencil to draw a large triangle with no sides or angles the same.

Use a ruler to find and mark the mid-point of each side.

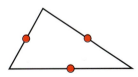

Use your ruler and a pencil to join the mid-points to make another triangle. Write down what you notice about:

a) the lengths of the sides of the two triangles

b) the angles of the two triangles.

- Distinguish between and estimate the size of acute, obtuse and reflex angles
- Use a protractor to measure acute, obtuse and reflex angles to the nearest degree

Key words
acute
right
obtuse
reflex
protractor

An angle is a measure of turn and is usually measured in degrees, or ° for short.

An **acute** angle is smaller than 90°.

A **right** angle is exactly 90°.

An **obtuse** angle is between 90° and 180°.

A **reflex** angle is greater than 180° but smaller than 360°.

We can use a **protractor** to measure angles.

Read off the size of the angle from the correct scale.
Angle ∠**ABC** = 25°

A protractor has two scales: a **clockwise** and an **anticlockwise** one. It is important to use the correct scale when measuring angles.

The vertex **B** is at the centre of the protractor.

The line segment **BA** is on the 0° line of the protractor.

Sometimes it is necessary to extend the lines before you can measure an angle.

Example   a) Give the name for this angle.
            b) Measure the angle.

a) Reflex

It is greater than 180° but smaller than 360°.

b)

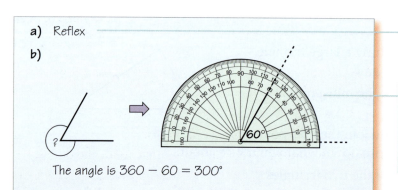

Extend the lines. Since we know that angles at a point add up to 360°, we can measure the size of the acute angle and then subtract this from 360°.

The angle is 360 − 60 = 300°

Exercise 2.2

1 Arrange these lettered angles in order of size, smallest first.

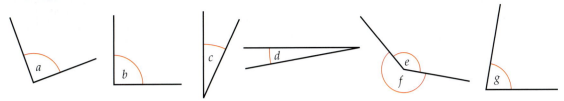

2 **i)** Give the names for each of these angles.
   ii) Estimate the sizes of each of these angles.

a)   **b)**   **c)**   **d)**   **e)**   **f)**

3 Play with a partner. Take turns to draw an angle. Both of you write down an estimate for the size of the angle, then measure it. Whoever is the closer to the estimate gains one point. If the estimate is within 5° of the actual size you get two points. Repeat five times each to find the overall winner.

4 Draw a rectangle with sides of 6 cm and 10 cm, using a ruler and set square.

Draw in one of the diagonals and measure the angles indicated on this diagram.

Write down what you notice about:

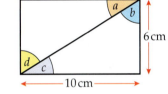

a) angles a and b **b)** angles c and d
c) angles a and c **d)** angles b and d?

5 Draw two straight lines that cross like this:

Measure the size of the angles on your diagram.

What do you notice? Repeat for another set of straight lines that cross. What do you notice?

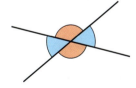

6 Draw any triangle and extend each of the sides as shown here.

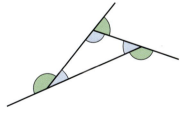

> The 'outside' angles are called **exterior** angles.

a) Measure each of the six angles indicated, and mark their sizes on the diagram.
b) Do the angles on each straight line add up to 180°?
c) Add together the angles inside the triangle. What do they add up to?
d) What do the three shaded angles outside the triangle add up to?
e) What do you notice about the inside angles? What do you notice about the outside angles?
f) Repeat parts **a)** to **e)** for another triangle.

Measuring and drawing angles

- Measure and draw lines to the nearest millimetre
- Measure and draw angles, including reflex angles, to the nearest degree

Key words
line segment
ruler
protractor
acute
right
obtuse
reflex

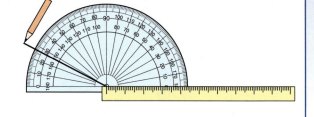

A **line segment** is a line with a fixed length and two end-points.

We can use a **ruler** and a **protractor** to measure and draw line segments and angles accurately.

Before drawing an angle, you need to think about what it will look like.

To help you draw it, first decide whether it is an **acute** angle, a **right** angle, an **obtuse** angle or a **reflex** angle.

For example, 153° is an **obtuse** angle.

This means that it is more than 90° ⌐ but less than 180°.

It will look something like this:

Example Draw a line segment AB of length 3.4 cm. Use a protractor to draw a reflex angle of 315° at the point B.

1)

A •————— 3.4 cm —————• B

Draw a line and mark the point A. Measure 3.4 cm from A and mark the point B. Remember to use a sharp pencil.

2)

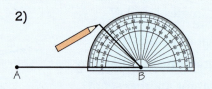

To draw an angle of 315° with an 180° protractor, place the protractor so that the 0° line is on AB. 360 − 315 = 45° so mark an angle of 45° using the scale that goes clockwise.

3)

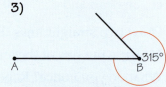

Take the protractor away and join the mark to the point B. Mark the angle of 315°.

Exercise 2.3

1 Sketch, then draw the following acute angles:
  a) 45° **b)** 63°
  c) 22° **d)** 85°

2 Sketch, then draw the following obtuse angles:
  a) 165° **b)** 143°
  c) 102° **d)** 137°

3 Sketch, then draw the following reflex angles:
   a) 310°   **b)** 280°
   c) 220°   **d)** 202°

4 Draw a line 5.6 cm long with an angle of 142° at one end.

5 Draw a line segment AB of length 6.4 cm.
   Draw an angle of 55° at A with a long line.
   Measure 3.2 cm along this line and mark the point C.

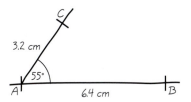

Here is a sketch of what your diagram should look like.

6 Draw a square with sides of length 4.8 cm, using a set square and a ruler.
   Draw in a diagonal and measure its length.

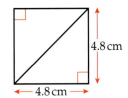

7 a) Draw a line segment AB of length 5.7 cm.
   Draw an angle of 43° at A with a long line.
   Measure 7.3 cm along this line and mark the point C.
   Join the point C to the point B to make a triangle.
   Measure the third side of your triangle and the other angles.

   b)

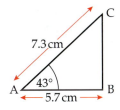

   Can you draw a different triangle with sides 5.7 cm and 7.3 cm and an angle of 43°?
   If so, measure the length of the third side and the other angles.
   Compare your triangles with a partner.

⊕ Be able to draw triangles using a ruler and protractor

Key words
draw
sides
angles
included angle
sketch
vertex

We can use a ruler and protractor to **draw** triangles accurately.

To do this, we need information about their **sides** and **angles** .

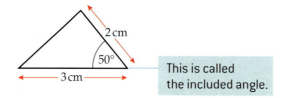

This is called the included angle.

If we know the lengths of two sides of the triangle and the size of the **included angle** (the angle in between them), there is only one possible triangle that we can draw. We call this drawing a triangle using SAS (side, angle, side) information.

It is often helpful to draw a **sketch** of the triangle, marking on all the information that we know, before we draw it accurately.

Example   Draw a triangle ABC with sides AB = 4 cm and AC = 2.4 cm where ∠BAC = 40°.

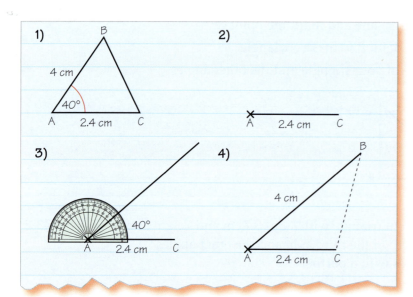

1) Draw a sketch first and mark on all the measurements.
2) Draw a base line AC of 2.4 cm. Mark a **vertex** for the angle at A.

3) Draw an angle of 40° with a long line.
4) Mark a point B that is 4 cm away from A. Complete the triangle by joining BC.

Exercise 2.4

1 Draw the shapes sketched below accurately. Measure the angle with the star in each one and write it down.

a)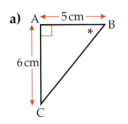

b)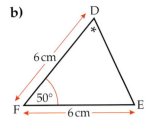

c)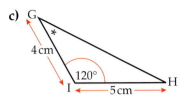

d)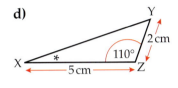

e)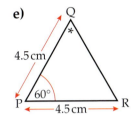

f)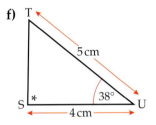

2 Draw a triangle with sides of length 5 cm and 6 cm, where the included angle is 60°.
Remember to draw a sketch first.
Measure and write down the length of the third side.

3 Draw a triangle with sides of length 5.2 cm and 6.4 cm, where the included angle is 48°.
Remember to draw a sketch first.
Measure the size of the other two angles.
Add up the three angles. Is your answer close to 180°?

4 Draw a triangle ABC where AB = 45 mm, AC = 80 mm and ∠BAC = 50°.
Measure and write down the length of BC.

> ∠BAC is the angle at A.

5 Draw a triangle DEF where DE = 52 mm, EF = 61 mm and ∠DEF = 160°.
Measure and write down the length of DF.
Measure the size of the other two angles.
Add up the three angles. Is your answer close to 180°?

> ∠DEF is the angle at E.

6 Draw a line segment AB of length 7.5 cm.
Draw an angle of 74° at A with a long line.
Draw an angle of 50° at B and extend the line until it meets the other line and makes a triangle. Mark the point where the lines meet as C.
Measure the sides AC and BC of your triangle and the angle ∠ACB. Can you draw a different triangle with a side of 7.5 cm and angles of 74° and 50°? If so, measure the length of the other two sides and the other angle. Compare your triangles with a partner.

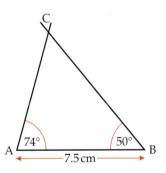

Parallel and perpendicular lines

Key words
parallel
perpendicular
right angles
vertically
   opposite

◈ Recognise and draw parallel and perpendicular lines using a ruler and set square

◈ Recognise vertically opposite angles

Parallel lines are straight lines that never meet or cross.

They are always the same distance apart.

We use arrows to show when lines are parallel.

You can use a set square and ruler to draw parallel lines.

Draw the line AB. Move the set square along the ruler then draw the line CD.

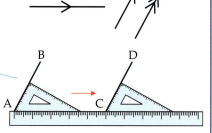

Perpendicular lines cross or meet each other at **right angles** (90°).

You can use a set square and ruler to draw perpendicular lines.

Draw the line EF. Place the set square onto the line and draw a perpendicular line GH.

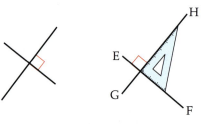

Vertically opposite angles are equal.

b and d are vertically opposite angles. $a = c$ and $b = d$

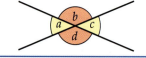

Example   Calculate the size of the lettered angles. Give reasons for your answers.

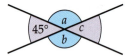

$c = 45°$ because vertically opposite angles are equal.

$a = 180° - 45° = 135°$ because angles on a straight line add up to 180°.

$b = 135°$ because vertically opposite angles are equal.

You could also say that $b = 180° - 45° = 135°$ because angles on a straight line add up to 180°.

Exercise 2.5

1 Name the pairs of lines that are **i)** parallel **ii)** perpendicular.

a)   b)   c)   d)   e)

f)    g)   h)   i)

2 Which is which? Copy these sentences and fill in the gaps.

    a) …… and …… are parallel.

    b) …… and …… are perpendicular to each other.

    c) …… is vertical.

    d) …… and …… are horizontal.

    There may be more than one answer.

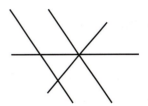

3 Draw these diagrams accurately using a ruler and set square.

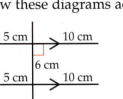 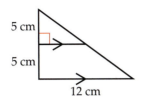

4 Copy this diagram.
Colour each pair of vertically opposite angles,
using a different colour for each pair.

5 Calculate (do not measure) the lettered angles. Give reasons for your answers.

    a)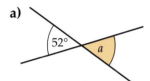

    b)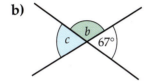

    c)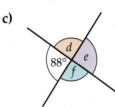

6 A rectangle measures 6 cm by 8 cm. Use a ruler and set square to draw the rectangle accurately and label the parallel and perpendicular lines. Measure the lengths of the diagonals.

7 Trace this diagram. Measure the lettered angles (you may need to extend the lines).
Check that angles on a straight line add up
to 180° and that the angles in the triangle
add up to 180°.
What do you notice about:

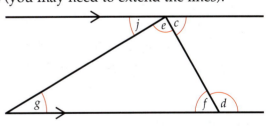

    a) angles *g* and *j*

    b) angles *c* and *f*?

Investigation

8 Here is a picture of a tangram.

These short lines mean the line
segments are the same length.

    Use a set square and ruler to
draw a tangram like this accurately.
Cut out the pieces. Investigate shapes you
can make, which have:

    a) at least one pair of parallel sides   **b)** at least one right angle.

    Make as many shapes as you can and draw them on plain paper.

2.6 Angle calculations

⊕ **Know the sum of angles in a triangle**

Key words
vertically
  opposite
equilateral
isosceles
right-angled

 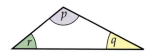 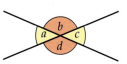

$a + b = 180°$
Angles that meet at a point on a straight line add up to **180°**.

$p + q + r = 180°$
The three angles of any triangle add up to **180°**.

$a = c$ and $b = d$
Vertically opposite angles are equal.

An **equilateral** triangle has three equal sides and three equal angles.

An **isosceles** triangle has two equal sides and two equal angles.

A **right-angled** triangle has one angle of 90°.

Example 1 Calculate the size of angles a, b and c.
Give reasons for your answers.

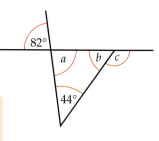

$a = 82°$ because it is vertically opposite an angle of 82°.
$b = 180° - 82° - 44°$
  $= 54°$ because the angles of a triangle add up to 180°.
$c = 180° - 54°$
  $= 126°$ because the angles on a straight line add up to 180°.

Example 2 Calculate the size of ∠CAB.
Give reasons for each step of your working.

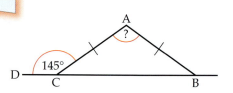

$∠ACB = 180° - 145°$
  $= 35°$ because the angles on a straight line add up to 180°.
$∠ABC = ∠ACB = 35°$ because the base angles of an isosceles
  triangle are equal.
$∠CAB = 180° - 35° - 35°$
  $= 110°$ because the angles of a triangle add up to 180°.

Exercise 2.6

1 Calculate the size of the lettered angles. Give reasons for your answers.

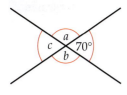 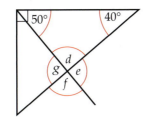 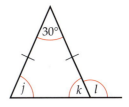

2 Calculate the size of ∠ABC in each of these diagrams.
Give reasons for each step of your working.

a) 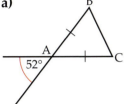 **b)** 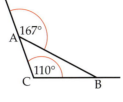 **c)**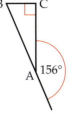

3 Calculate the size of the lettered angles. Give reasons for each step of your working.

a) 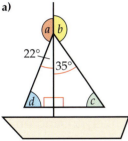 **b)**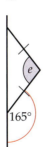

4 Together, the pink triangle and the grey triangle make one big triangle.
Calculate the size of the angles marked *a*, *b* and *c*.
Write down the size of the angles of the big triangle, outlined in blue. What do they add up to?

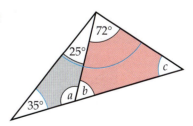

5 Sketch an isosceles triangle ABC.
Label your diagram to show that AB = BC.
Mark ∠ACB = 50°.
Calculate the size of the other angles in the triangle.

Probability using words

⊕ Describe probability using words

Key words
event
probability scale
random

You can describe the probability or (chance) of an **event** happening in words. This **probability scale** shows some of them.

```
0                                      1
└──────────────┴──────────────┘
impossible        even        certain
```

A **random** event is one that can't be predicted. Throwing a dice is one way of producing random numbers.

Example 1   Jack chooses a counter at random from this jar. Draw a probability scale showing the approximate chance of each colour being chosen.

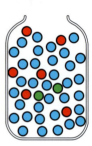

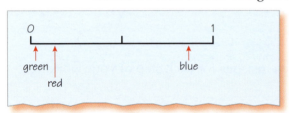

Example 2

a)   Colour this spinner so it is less likely that it will land on blue than yellow.

b)   Colour this spinner so there is an equal chance of landing on green and yellow but it is most likely to land on blue.

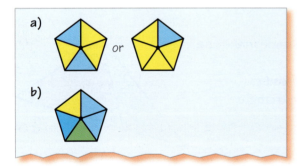

a)

or

b)

Exercise 3.1

1 Here is the net of a dice.
   a) Which number is most likely to be thrown?
   b) Which number is least likely to be thrown?
   c) Which number has an even chance of being thrown?
   d) It is impossible to throw some numbers between 1 and 6 using this dice. Which ones are they?

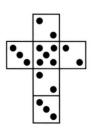

2 50 raffle tickets are sold, numbered from 1 to 50. A ticket is chosen at random, so each has the same chance of being chosen.

Match each event to the chance of it being chosen. One has already been done for you.

A The number 16
B The number 75
C An odd number
D A number less than 51
E A number bigger than 10

1 certain
2 likely
3 impossible
4 unlikely
5 even

E2 (It is likely that a number bigger than 10 will be chosen.)

3 A counter is chosen at random from each jar.

Draw a probability scale for each jar to show the approximate chance of a yellow, blue, red and green counter being chosen.

Use Example 1 to help you.

a) 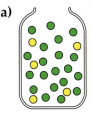 b)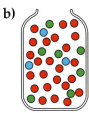

4 You need blue, green and red colouring pencils. Copy this spinner twice.

a) Colour the spinner so it is impossible for it to land on blue and more likely to land on green than red.

b) Colour the spinner so it is most likely to land on blue but green and red have an equal chance as each other.

5 Here are the nets of three different dice.

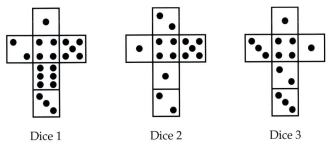

Dice 1   Dice 2   Dice 3

Which dice would you choose so:

a) there is an equal chance of throwing an odd or even number
b) it is certain to throw a number less than 5
c) there is an equal chance of throwing a 5 or 6
d) it is impossible to throw a 3?

There are two answers to this question.

6 Eliza says that she will either meet a zebra on her way to school or she will not, so there is an even chance of meeting a zebra. Is Eliza right? Explain your answer.

Investigation

7 Think of an event which is:
a) impossible   **b)** certain
c) has an even chance of happening
d) has a chance between even and certain of happening
e) has a chance between impossible and even of happening.

Place each of your events on the probability scale.

0 ⊢——————⊣ 1
impossible   even   certain

Probability using numbers

◈ Describe probability using fractions, decimals and percentages

A probability can be written as a fraction or decimal. This value must be from 0 to 1. A probability can also be written as a percentage.

A **random** choice is one that can't be predicted. The **theoretical** probability of choosing a red counter at random is $\frac{2}{5}$. This is because there are two red counters out of a total of five, and they each have the same chance of being chosen.

$\frac{2}{5} = 0.4 = 40\%$, so the probability can also be written as 0.4 or 40%.

This probability should not be written as '2 in 5 chance', '2 : 5' or '2 out of 5'.

Example   A bag contains ten Scrabble tiles:

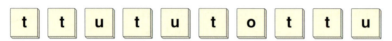

A tile is taken out at random. Find the probability that the letter is a:
a) 't'   b) 'o'   c) 'u'
d) a vowel   e) 'i'   f) 't', 'o' or 'u'
Give each answer as a fraction, decimal and percentage.
Draw a probability scale for each of these events.

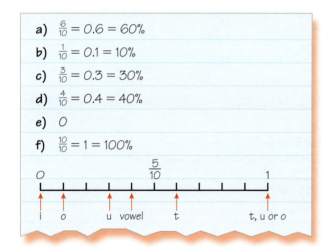

a) $\frac{6}{10} = 0.6 = 60\%$

b) $\frac{1}{10} = 0.1 = 10\%$

c) $\frac{3}{10} = 0.3 = 30\%$

d) $\frac{4}{10} = 0.4 = 40\%$

e) 0

f) $\frac{10}{10} = 1 = 100\%$

Exercise 3.2

1 Sam has ten pens in his pencil case. Three are blue, two are red and five are black. He takes a pen out at random. What is the probability that the pen is:
a) black   b) blue   c) red
d) green   e) black, blue or red?
f) Draw a probability scale showing each of these events.

Use the Example to help you.

2 A dice is made from this net.
The dice is then thrown.

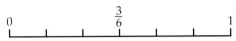

a) What is the probability that the dice shows:

   i) 4   **ii)** 3   **iii)** 1 or 4
   iv) an even number   **v)** 6   **vi)** 1, 3 or 4?

b) Copy and complete the probability scale to show your answers.

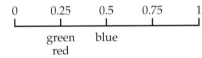

3 In a class there are 14 boys and 11 girls. The teacher chooses a pupil at random to hand out the exercise books. What is the probability that the teacher chooses a boy?

4 A school sells 100 raffle tickets. Tickets numbered from 1 to 60 are blue and from 61 to 100 are red. A ticket is drawn at random. What is the probability that this ticket is:

a) blue   **b)** red

c) an even number   **d)** a number that ends in zero

e) the number 23   **f)** blue and a number greater than 60.

5 Draw a spinner that could be described by this probability scale.

```
0        0.25      0.5       0.75      1
|_____|_____|_____|_____|
         green     blue
          red
```

6 Ruby has 4 sweets in a bag. She always chooses a sweet without looking. The probability of her choosing a mint is $\frac{1}{2}$. Ruby eats one of the mint sweets. What is the probability that the next sweet she chooses at random is also a mint?

> You could use four counters to help.

7 Mrs Smith has five keys in her bag. She chooses a key at random.

a) What is the probability that she chooses the key that she needs?

b) The key Mrs Smith chooses does not open her door. What is the probability that the next key she chooses opens her door if:

   i) she puts the key she has already used back in her bag first
   ii) she does not put the key she has already used back in her bag.

You can try this out using five different coloured counters.

8 Mrs Smith also has a pair of red gloves and a pair of blue gloves in her bag.

> You could use two red and two blue counters to help.

a) What is the probability that the first glove she chooses at random is blue?

b) What is the greatest number of gloves she needs to take out from her bag to give a pair of the same colour?

Investigation

9 Use the idea from **Q5** to draw three different spinners and their probability scales. Use fractions rather than decimals for your probability scale.

3.3 Possible outcomes

● Record outcomes using diagrams and tables

Key words
outcome
event
sample space diagram

You can show possible **outcomes** of an **event** using a table or a **sample space diagram**. For example, for the event 'tossing a coin', the outcome is either a *tail* or a *head*.

Example 1   Find all possible outcomes when two coins are tossed.

Coin 1 / Coin 2	Head	Tail
Head	Head, head	Tail, head
Tail	Head, tail	Tail, tail

There are four possible outcomes.

Example 2   Ali, Ben and Colin go to the fair. Only two of the boys can go on a ride at the same time.

a)   What are all the possible outcomes?

b)   What is the probability that Ali and Colin go on the ride together?

a)   Ali and Ben; Ali and Colin; Ben and Colin.

b) $\frac{1}{3}$

There are three possible outcomes and only one of them is Ali and Colin.

Exercise 3.3

1 A shop sells chocolate, vanilla or strawberry ice cream. Sangheeta would like to try two different flavours. What combinations could she have?

2 Another shop sells ice creams and a choice of topping.

a)   Mark wants one flavour of ice cream with a topping. What are his choices?

b)   Louise wants two different flavours of ice cream and a topping. What are her choices?

Ice cream
Mint
Vanilla
Banana

Topping
Sauce
or
Flake

3 Copy and complete the table to show all the possible outcomes when a dice is thrown and a coin is spun.

Dice / Coin	1	2	3	4	5	6
Head	1, head					
Tail						

What is the probability of getting 4, *tail*?

4 Luke has number cards 1, 3, 5. Adam has number cards 2, 4, 6. They each choose a card at random and add the two numbers together. Copy and complete the table to show all of the possible outcomes.

+	1	3	5
2	3		
4			
6			

  a) Use your table to find the probability that the total will be:

      i) 5   **ii)** less than 6

      iii) greater than 6   **iv)** an even number.

  b) Complete the table to show all of the possible outcomes if the numbers are multiplied together. Use your table to find the probability that the total will be:

×	1	3	5
2	2		
4			
6			

      i) 5   **ii)** less than 6

      iii) greater than 6   **iv)** an even number.

      Are your answers the same for each table?

5 Pran spins these two spinners. He adds together the numbers that each spinner lands on. Draw a table to show all of the different possible totals.

Is the probability of an **even** total the same as the probability of an **odd** total? Explain your answer.

6 Matt has the number cards 4, 5 and 6.

  a) Which 2-digit numbers can he make?

  b) What is the probability that a 2-digit number chosen at random is:

      i) even   **ii)** less than 50   **iii)** not less than 50   **iv)** divisible by 5?

> 64 is an example of a 2-digit number.

Investigation

7 Copy and complete these two tables showing the outcomes when two dice are thrown.

  a) Fill in the boxes by adding the numbers on the dice.

+	1	2	3	4	5	6
1	2	3				
2						
3			7			
4						
5						
6						

  b) Fill in the boxes by subtracting the smallest number from the largest.

−	1	2	3	4	5	6
1	0					
2			1			
3						
4					2	
5						
6		4				

> You will need these results in your next lesson.

Tallies and frequency tables

⊕ Collect data from a simple experiment and record it in a frequency table

A frequency table is a way of sorting data into groups. It is often quicker and easier to record data using a **frequency** table.

Example   Eric asks some pupils in his school how many pieces of homework they got yesterday.

0	1	1	1	2	1	1	4	1	2	2	1
1	1	3	1	2	1	3	1	0	1	0	2

Draw a tally and frequency table for this information.

Number of pieces of homework	0	1	2	3	4
Tally	|||	ＪＨＴ ＪＨＴ |||	ＪＨＴ	||		
Frequency	3	13	5	2	1

Remember that tally marks are grouped in 5's.

Exercise 3.4

1 A coach stops at a café.

Here is the passengers' order:

a) How many adults ordered Orange juice?

b) How many Coffees were ordered?

c) How many more adults than children chose water?

d) How many drinks were ordered in total?

Drink	Adult	Child
Tea	ＪＨＴ ＪＨＴ	/
Coffee	///	ＪＨＴ ＪＨＴ //
Orange Juice	ＪＨＴ	ＪＨＴ //
Water	ＪＨＴ ////	///

2 Mike asks pupils in his form group which lesson is their favourite.

science	P.E.	maths	English	science	P.E.
English	maths	English	science	P.E.	English
P.E.	science	maths	P.E.	English	P.E.

a) Draw a tally and frequency table for his information.

b) Which lesson is the most popular?

3 Three pupils record in different ways the types of vehicle that pass their school.

Sue

bus car lorry
car bike

Christine

b   c   l   c   b

Matt

bus	/
car	//
lorry	/
bike	/

a) What problem may Sue have when she does her traffic survey?

b) What problem may Christine have when she does her traffic survey?

c) Why is Matt's way the best to record this information?

4 This activity requires a coin.

Sam throws two coins 60 times recording the number of *tails* shown for each throw.

Copy the table then carry out Sam's experiment, recording the results in your table.

'I will get about 20 "no tails", about 20 "1 tail" and about 20 "2 tails'.

Number of *tails*	0	1	2
Tally			
Frequency			

Is Sam correct? Explain your answer.

Investigation

5 Requires children's building blocks.

Ben thinks that if a building block is thrown, it will not land on each face with the same frequency.

Carry out an experiment with different sizes of building block to test his hypothesis.

Use a table with these headings to record your results:

Size of face	Small	Medium	Large

Use your results to decide if Ben is correct.

6 Use the results from your investigation in your last maths lesson.

Make a tally chart for each number in the two tables.

		0	1	2	3	4	5	6	7	8	9	10	11	12
Adding the numbers	Tally													
	Frequency	0	0											
Subtracting the numbers	Tally													
	Frequency							0	0	0	0	0	0	0	
Total frequency														

Rhys says the number 4 has the highest probability of being a result.

Use your completed table to say whether Rhys is right or wrong.

Estimating probability

◈ Estimate probability from experiments.

When we carry out an experiment we can use the results to find the **estimated** probability of each outcome. For example, if a coin is thrown 10 times and 6 *tails* are recorded, the estimated probability of the coin landing on *tails* is $\frac{6}{10}$ or 0.6.

Example 1 A sweet machine dispenses fruit lollipops.
The fruit is randomly selected by the machine.

Di records the frequency of each fruit.

Fruit	Orange	Cherry	Lemon	Strawberry
Frequency	2	4	8	6

a) What is the estimated probability of each fruit?
b) Which fruit is the machine most likely to select next?

a)

Fruit	Orange	Cherry	Lemon	Strawberry
Frequency	2	4	8	6
Estimated probability	$\frac{2}{20} = 0.1$	$\frac{4}{20} = 0.2$	$\frac{8}{20} = 0.4$	$\frac{6}{20} = 0.3$

$2 \div 20 = 0.1$

b) Lemon

Example 2 Ellen plays lucky dip at the fair. Each go costs 10p, and she has a chance of winning 10p, winning 20p, or losing.

	Win 10p	Win 20p	Lose
Frequency	5	1	4
Estimated probability			

a) Copy and complete the table showing the estimated probabilities of winning and losing.
b) How much did Ellen spend on the lucky dip?
c) How much did Ellen win?
d) How much in total did Ellen lose playing the game?

a)

	Win 10p	Win 20p	Lose
Frequency	5	1	4
Estimated probability	$\frac{5}{10} = \frac{1}{2}$	$\frac{1}{10}$	$\frac{4}{10} = \frac{2}{5}$

b) Ellen spent $10 \times 10p = £1$
c) Ellen won $5 \times 10p + 1 \times 20p = 50p + 20p = 70p$
d) Ellen lost $£1 - 70p = 30p$

Exercise 3.5

1 Three dice are thrown.

Number of dice showing the same outcome	3	2	0
Frequency	2	20	28
Estimated probability			

a) How many times were the three dice thrown?

b) Copy and complete the table showing the estimated probabilities.

2 Ten counters are placed in a bag. Bindi chooses a counter at random, notes the colour then puts the counter back. She repeats this 20 times.

Colour of counter	Blue	Red	Yellow
Frequency	11	6	3
Estimated probability			

a) Copy and complete the table to find the estimated probability of choosing each colour.

b) How many of each colour of counter are likely to be in the bag?

> Remember that there are 10 counters. There are two likely answers.

3 Dan plays a game at a school fair. Here is what he won and lost.

Outcome	Win 20p	Win 50p	Lose
Frequency	3	1	6
Estimated probability			

> Use Example 2 to help.

a) How many times did Dan play the game?

b) Copy and complete the table for the estimated probabilities.

c) It costs 10p to play each time. How much did Dan spend?

d) How much money did Dan win?

e) Did Dan lose or win money overall? How much?

4 This question requires a dice.
Copy this table:

Number of throws	1	2	3	4	5	6
Tally						
Frequency						
Estimated probability						

> If you throw 2, 2, 5 it has taken you **3** throws to total 6 or more, so you would put a tally mark in the **3** column.

Throw the dice, adding up the numbers thrown until you reach a total of 6 or more. Record in a table the **number of throws** needed to get this total. Repeat this experiment 50 times. Complete the table by finding the estimated probabilities for each number of throws. Write each answer as a fraction and a decimal. For example, if it took you 3 throws only **once** out of the 50 tries, the estimated probability would be $\frac{1}{50}$ or 0.02.

Comparing probability

⊕ Compare theoretical and experimental probabilities

If you toss a fair coin, there is an equal chance of getting a *head* or a *tail*. But if the coin was **biased** (or unfair), there would be a greater chance of either *heads* or *tails*. We know that the probability of throwing a '2' on a fair dice is $\frac{1}{6}$. This is the **theoretical** probability. When we actually throw the dice a number of times we can use our results to calculate the **experimental** probability.

We do not expect the values for the theoretical and experimental probabilities to always be the same.

Example Caitlin throws two coins and counts the number of *tails* showing for each throw.

a) Find the theoretical probability for each outcome.

b) Caitlin throws the two coins 100 times. Here are her results:

Number of *tails*	0	1	2
Frequency	23	49	28

Find the experimental probability for each outcome.

c) Compare the theoretical and estimated probabilities, to find out if the coin is fair.

a)
Number of tails	0	1	2
Outcome	head, head	head, tail / tail, head	tail, tail
Theoretical probability	$\frac{1}{4}$	$\frac{1}{4}+\frac{1}{4}=\frac{1}{2}$	$\frac{1}{4}$

There are four outcomes.

b) Experimental probability of 0 tails is $\frac{23}{100}$, of 1 tail is $\frac{49}{100}$ and of 2 tails is $\frac{28}{100}$.

c) $\frac{25}{100}$ is the same as $\frac{1}{4}$, and $\frac{50}{100}$ is the same as $\frac{1}{2}$, so the experimental and theoretical probabilities are quite close. The coin is fair.

Exercise 3.6

1 Jane puts 20 counters in a bag: 10 blue, 6 red and 4 orange. She chooses a counter at random, records its colour and returns it to the bag. She does this 20 times. Here are her results:

	Blue	Red	Orange
Experimental probability	$\frac{9}{20}$	$\frac{6}{20}$	$\frac{5}{20}$

a) What are the theoretical probabilities for each colour?

b) Jane says that she must have done the experiment wrong as the results for the theoretical probabilities are not the same as the experimental ones. Is Jane correct? Explain your answer.

2 Activity for two players. Requires counters and a bag.

Pupil A chooses 10 counters of a mixture of blue and red colours and places them in a bag, without Pupil B seeing them. Pupil B takes a counter from the bag, notes the colour and replaces it. Repeat this 10 times, then estimate the number of each colour. Next, repeat the experiment a total of 30 times, then 50 times. You can record your results in a table.

Number of counters taken from bag	10	30	50
Number of red counters chosen			
Estimated probability			
Number of blue counters chosen			
Estimated probability			

Empty the bag to see how many counters of each colour there are.

Calculate the theoretical probabilities then compare the theoretical probabilities with the experimental probabilities for 50 repeats of the experiment. Write a sentence about your results.

3 Requires square card and a cocktail stick.

Colour your spinner like the one shown but when you have made the spinner, do not place the cocktail stick in the centre of your spinner, so it is **biased** or not fair.

Spin your spinner 40 times, recording the colour it lands on.

Copy and complete the table. Calculate the probability of the spinner landing on each colour if it was not biased.

Colour	Red	Blue	White	Yellow
Tally				
Frequency				
Estimated probability				
Probability of landing on colour if not biased				

4 Requires random number tables or a random function on a calculator.

Charlie says that a number chosen at random has the same probability of being odd or even.

	Odd	Even
Tally		
Frequency		
Experimental probability		
Theoretical probability		

> 0 is an even number for this activity.

Use the random number table or the random function on a calculator to simulate 40 odd or even numbers, by recording if the last digit is odd or even.

For example the number 0.3072 would be recorded as even as it ends with a 2.

a) How many numbers out of 40 would you expect to be odd if Charlie is correct?

b) Copy and complete the table showing each type of probability.

Equivalent fractions

◈ Recognise equivalent fractions

◈ Simplify fractions by cancelling

Key words
equivalent fraction   cancel
numerator   simplify
denominator   simplest form

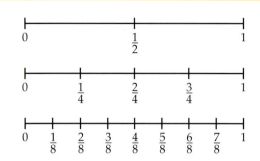

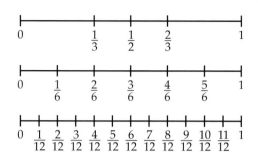

Equivalent sets of fractions are:

$\frac{1}{2}, \frac{2}{4}, \frac{4}{8}$ $\frac{1}{3}, \frac{2}{6}, \frac{4}{12}$ $\frac{3}{4}, \frac{6}{8}$

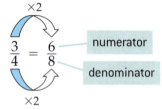

Equivalent fractions can be created by multiplying or dividing the **numerator** and **denominator** by the same number.

Dividing a numerator and denominator by the same number is called **cancelling**.

When a fraction cannot be cancelled it has been **simplified**, and is in its **simplest form** or lowest terms.

For example $\frac{8}{12} = \frac{4}{6} = \frac{2}{3}$

Example 1   Complete these equivalent fractions:

a) $\frac{2}{3} = \frac{50}{\square}$ b) $\frac{15}{20} = \frac{\square}{4}$

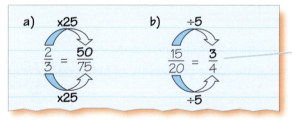

The denominator (20) has been divided by 5 to give 4, so the numerator must also be divided by 5.

Example 2   Simplify the following:

a) $\frac{30}{70}$ b) $\frac{16}{24}$ c) $\frac{28}{42}$

a) $\frac{30}{70} = \frac{3}{7}$ Divide numerator and denominator by 10.

b) $\frac{16}{24} = \frac{8}{12} = \frac{4}{6} = \frac{2}{3}$ Divide both numbers by 2 each time. We could have divided by 8 and got the answer in one step.

c) $\frac{28}{70} = \frac{14}{35} = \frac{2}{5}$ Divide by 2 and then by 7.

Exercise 4.1

1 Copy and complete these equivalent fractions.

÷2

a) $\frac{6}{10} = \frac{\square}{5}$

b) $\frac{10}{12} = \frac{5}{\square}$

c) $\frac{8}{16} = \frac{\square}{2}$

d) $\frac{15}{25} = \frac{3}{\square}$

÷2

2 Copy and complete these equivalent fractions.

×3

a) $\frac{2}{3} = \frac{\square}{9}$

b) $\frac{5}{8} = \frac{20}{\square}$

c) $\frac{4}{5} = \frac{12}{\square}$

d) $\frac{3}{4} = \frac{\square}{40}$

×3

3 Copy and complete the cancelling to find the simplest term.

÷10   ÷2

a) $\frac{60}{80} = \frac{6}{\square} = \frac{\square}{4}$

b) $\frac{16}{40} = \frac{\square}{20} = \frac{4}{\square} = \frac{\square}{5}$

÷10   ÷2

4

$\frac{6}{9}$ $\frac{3}{4}$ $\frac{6}{10}$ $\frac{6}{8}$ $\frac{10}{15}$ $\frac{2}{3}$

Find the fractions from the cloud which are equivalent to the following numbers.

a) $\frac{3}{5}$

b) $\frac{4}{6}$

c) $\frac{9}{12}$

> Try multiplying up or dividing down.

5 Cancel these fractions down to their lowest terms.

a) $\frac{3}{6}$
b) $\frac{5}{20}$
c) $\frac{4}{12}$
d) $\frac{2}{10}$
e) $\frac{5}{15}$
f) $\frac{4}{10}$

g) $\frac{4}{16}$
h) $\frac{6}{24}$
i) $\frac{30}{50}$
j) $\frac{25}{50}$
k) $\frac{46}{60}$
l) $\frac{100}{150}$

6 Find the odd one out by writing these fractions in their simplest form.

a) $\frac{4}{6}$ $\frac{6}{9}$ $\frac{7}{14}$ $\frac{8}{12}$ $\frac{10}{15}$

b) $\frac{9}{15}$ $\frac{20}{25}$ $\frac{6}{10}$ $\frac{12}{20}$ $\frac{30}{50}$

c) $\frac{15}{20}$ $\frac{60}{80}$ $\frac{10}{16}$ $\frac{9}{12}$ $\frac{18}{24}$

7 a) Which of these fractions **cannot** be simplified?

$\frac{2}{3}$ $\frac{6}{8}$ $\frac{15}{30}$ $\frac{7}{12}$ $\frac{8}{10}$ $\frac{4}{9}$ $\frac{13}{39}$ $\frac{3}{5}$

b) How can you tell when a fraction is in its simplest form?

Fractions and decimals

◈ Convert decimals to fractions

◈ Convert fractions to decimals with and without a calculator

Key words
decimal
fraction
equivalent fraction
denominator
numerator

To convert a **decimal** to a **fraction**, write it as a number of tenths, or hundredths, or thousandths, and then simplify it by cancelling.

For example: $0.8 = \frac{8}{10} = \frac{4}{5}$

$0.64 = \frac{64}{100} = \frac{16}{25}$

units	tenths	hundredths	thousandths
0	8		
0	6	4	

To convert a fraction to a decimal, either:

a) Find an **equivalent fraction** with a **denominator** of 10, 100, 1000, …

For example: $\frac{7}{20} = \frac{35}{100}$

$= 0.35$

> Multiply top and bottom by 5.

or b) Divide the **numerator** by the denominator using a calculator.

For example: $\frac{7}{8} = 0.875$

> A fraction is the same as a division, so $\frac{7}{8}$ means $7 \div 8$.

[7] [÷] [8] [=]

Example 1 Convert these decimals into fractions in their lowest terms.

a) 0.4   b) 0.48   c) 0.125

a) $0.4 = \frac{4}{10} = \frac{2}{5}$

> One decimal place shows it is 4 tenths.

b) $0.48 = \frac{48}{100}$

$\frac{48}{100} = \frac{24}{50} = \frac{12}{25}$

> 0.48 means 4 tenths and 8 hundredths. 4 tenths is 40 hundredths, so in total it is 48 hundredths.

c) $0.125 = \frac{125}{1000}$

$\frac{125}{1000} = \frac{5}{40} = \frac{1}{8}$

> 0.125 means 1 tenth, 2 hundredths and 5 thousandths (100 + 20 + 5 thousandths), or 125 thousandths.

Example 2 Convert these fractions into decimals.

a) $\frac{3}{5}$ b) $\frac{7}{20}$ c) $\frac{3}{16}$

a) $\frac{3}{5} = \frac{6}{10} = 0.6$

> Multiplying top and bottom by 2 we can change the fifths into tenths.

b) $\frac{7}{20} = \frac{35}{100} = 0.35$

> Multiplying top and bottom by 5 we can change 20ths into 100ths.

c) $\frac{3}{16} = 0.1875$

> We cannot easily change 16ths into 100ths – so we divide.

Exercise 4.2

1 Convert these decimals into fractions. Give your answers in their simplest form.

    a) 0.3   **b)** 0.6   **c)** 0.1

    d) 0.8   **e)** 0.5

2 Convert these decimals into fractions. Give your answers in their simplest form.

    a) 0.25   **b)** 0.72   **c)** 0.36

    d) 0.45   **e)** 0.88

3 Convert these fractions into decimals.

    a) $\frac{4}{5}$ **b)** $\frac{11}{20}$ **c)** $\frac{3}{25}$

    d) $\frac{3}{4}$ **e)** $\frac{7}{50}$

4 Convert these fractions into decimals.

    a) $\frac{7}{8}$ **b)** $\frac{2}{5}$ **c)** $\frac{5}{16}$

    d) $\frac{6}{15}$ **e)** $\frac{9}{75}$

5 Convert these decimals into fractions in their simplest form.

    a) 0.7   **b)** 0.32   **c)** 0.375

    d) 0.65   **e)** 0.225

6 Use a set of 1–10 number cards. Shuffle them and deal out five cards. Use your cards to fill in these gaps:

    Now convert each fraction into a decimal.

7 Use a set of 0–9 digit cards. Shuffle and place 1, 2, or 3 cards to make a decimal, for example 0.125 or 0.38.

    Now convert your decimal into a fraction in its lowest terms.

Investigation

8 **a)** Convert $\frac{1}{3}$ and $\frac{2}{3}$ into decimals. What do you notice?

     b) Convert $\frac{1}{9}, \frac{2}{9}, \dots \frac{8}{9}$ into decimals. What do you notice?

Ordering fractions

⊕ Order fractions by changing them to equivalent fractions with a common denominator

⊕ Order fractions using a diagram

There are two methods for ordering fractions.

- **Method A:**

  Convert them to **equivalent fractions** with a **common denominator** (the same denominator).

  To put these fractions $\frac{4}{5}, \frac{1}{2}, \frac{17}{20}$ and $\frac{7}{10}$ in order, find the common denominator. This is the smallest number that all of the denominators 5, 2, 20 and 10 will divide into, i.e. 20. Now convert the fractions to equivalent 20ths:

$\frac{4}{5} = \frac{16}{20}$	Multiply top and bottom by 4.	$\frac{1}{2} = \frac{10}{20}$	Multiply top and bottom by 10.
$\frac{17}{20}$	It already has a denominator of 20.	$\frac{7}{10} = \frac{14}{20}$	Multiply top and bottom by 2.

  The order (from lowest to highest) is now easily seen, i.e. $\frac{1}{2}, \frac{7}{10}, \frac{4}{5}, \frac{17}{20}$.

- **Method B:**

  Use a diagram.

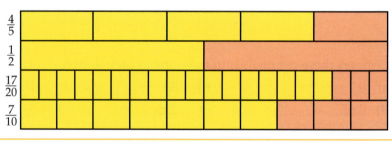

Example 1 By finding equivalent fractions with the same denominator, decide which is greater: $\frac{2}{3}$ or $\frac{3}{4}$

The smallest number that 3 and 4 will divide into is 12, so this will be the common denominator.

$\frac{2}{3} = \frac{8}{12}$ $\frac{3}{4} = \frac{9}{12}$

$\frac{3}{4} > \frac{2}{3}$

$\frac{9}{12}$ is greater than $\frac{8}{12}$.

Example 2 Using a diagram, place these fractions in order from smallest to largest.

$\frac{3}{5}$ $\frac{1}{20}$ $\frac{5}{8}$ $\frac{7}{10}$

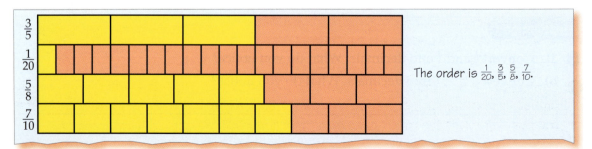

The order is $\frac{1}{20}, \frac{3}{5}, \frac{5}{8}, \frac{7}{10}$.

Exercise 4.3

1 ├──┼──┼──┼──┼──┼──┼──┤

Copy this number line and mark on:

$\frac{1}{2}$ $\frac{1}{4}$ $\frac{1}{8}$ $\frac{3}{4}$ $\frac{5}{8}$ $\frac{7}{8}$ $\frac{3}{8}$

Use your line to place these fractions in order from smallest to largest.

$\frac{3}{8}$ $\frac{6}{8}$ $\frac{1}{2}$ $\frac{1}{4}$ $\frac{7}{8}$

2 By finding equivalent fractions with the same denominator decide which fraction is the larger.

a) $\frac{1}{4}$ or $\frac{1}{3}$

b) $\frac{3}{5}$ or $\frac{7}{10}$

c) $\frac{7}{12}$ or $\frac{1}{2}$

d) $\frac{2}{3}$ or $\frac{4}{9}$

e) $\frac{2}{7}$ or $\frac{5}{14}$

f) $\frac{6}{20}$ or $\frac{2}{10}$ or $\frac{3}{5}$

3 By drawing diagrams, or any other method, find which fraction is the smaller.

a) $\frac{3}{5}$ or $\frac{7}{10}$

b) $\frac{1}{4}$ or $\frac{3}{20}$

c) $\frac{3}{8}$ or $\frac{1}{2}$

d) $\frac{9}{24}$ or $\frac{2}{5}$

e) $\frac{5}{8}$ or $\frac{9}{15}$

f) $\frac{18}{60}$ or $\frac{5}{16}$

4 Place these fractions in order from smallest to largest.

a) $\frac{3}{4}$ $\frac{13}{20}$ $\frac{3}{5}$ $\frac{1}{2}$ $\frac{7}{10}$

b) $\frac{3}{10}$ $\frac{3}{8}$ $\frac{19}{50}$ $\frac{1}{4}$ $\frac{27}{100}$

c) $\frac{3}{4}$ $\frac{91}{100}$ $\frac{17}{20}$ $\frac{9}{10}$ $\frac{2}{5}$

d) $\frac{7}{10}$ $\frac{5}{16}$ $\frac{16}{20}$ $\frac{14}{50}$ $\frac{7}{8}$

e) $\frac{1}{3}$ $\frac{34}{100}$ $\frac{3}{7}$ $\frac{3}{10}$ $\frac{4}{9}$

5 True or False?

a) $\frac{2}{5}$ is greater than $\frac{3}{10}$

b) $\frac{3}{4}$ is greater than $\frac{8}{9}$

c) $\frac{1}{2}$ is smaller than $\frac{9}{20}$

d) $\frac{4}{9}$ is greater than $\frac{3}{8}$.

6 John spent $\frac{3}{4}$ of his pocket money on magazines and Sue spent $\frac{7}{10}$ of hers on magazines. Who spent the greatest fraction of their pocket money on magazines?

7 Ann spent $\frac{1}{8}$ of her Sunday watching TV. Peter spent $\frac{3}{16}$ of his day watching TV. Who watched the most TV?

8 Rani was sharing out some chocolates. He gave $\frac{1}{4}$ to Jan, $\frac{3}{8}$ to Liam and $\frac{5}{16}$ to Sarah. Who got the most chocolates?

9 Use a set of 1–10 number cards. Shuffle them and deal out five pairs of cards. For each pair place the smaller number as the numerator and the larger as the denominator of a fraction. Write the five fractions in order, smallest to largest. Reshuffle and repeat.

Adding and subtracting fractions

⊕ Add and subtract fractions by writing them with a common denominator

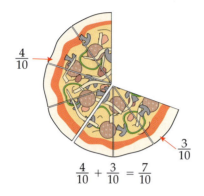

$$\frac{4}{10}$$

$$\frac{3}{10}$$

$$\frac{4}{10} + \frac{3}{10} = \frac{7}{10}$$

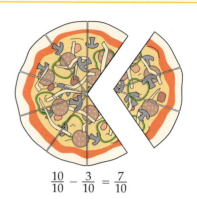

$$\frac{10}{10} - \frac{3}{10} = \frac{7}{10}$$

To add or subtract fractions:

When the denominators are the same, add or subtract the numerators.

When the denominators are different, first find **equivalent fractions** that have a **common denominator** .

To find the common denominator, find the smallest number that both denominators will divide into.

For example, the common denominator of $\frac{1}{2}$ and $\frac{3}{4}$ is 4.

$$\frac{1}{2} = \frac{2}{4}$$ Multiply top and bottom by 2.

So, $\frac{1}{2} + \frac{3}{4} = \frac{2}{4} + \frac{3}{4} = \frac{5}{4} = \frac{4}{4} + \frac{1}{4} = 1\frac{1}{4}$

$\frac{5}{4}$ is an **improper fraction** . $1\frac{1}{4}$ is a **mixed number** . Both are correct answers.

Example 1 Calculate:
$$\frac{3}{5} + \frac{4}{5} + \frac{1}{5}$$

The denominators are all the same, so add the numerators.

$$\frac{3+4+1}{5} = \frac{8}{5} = 1\frac{3}{5}$$

$\frac{5}{5} = 1$ whole

Example 2 Calculate:
$$\frac{3}{4} - \frac{1}{8}$$

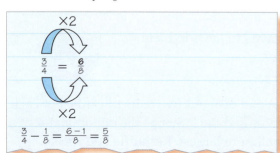

×2

$$\frac{3}{4} = \frac{6}{8}$$

×2

Change $\frac{3}{4}$ to eighths.

$$\frac{3}{4} - \frac{1}{8} = \frac{6-1}{8} = \frac{5}{8}$$

Exercise 4.4

1 Copy and complete these additions. The first one has been done for you.

a)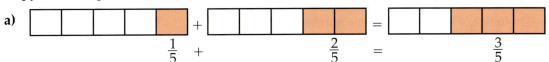

$$\frac{1}{5} \quad + \quad \frac{2}{5} \quad = \quad \frac{3}{5}$$

b)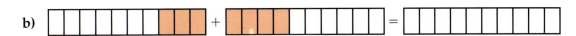

c)

2 Copy and complete these calculations.

a) $\frac{4}{9} + \frac{3}{9}$

b) $\frac{7}{12} - \frac{2}{12}$

c) $\frac{16}{20} - \frac{7}{20}$

d) $\frac{7}{15} + \frac{3}{15} + \frac{6}{15}$

3 What is the smallest common denominator of these numbers?

> The smallest common denominator of two numbers is the smallest number they will both divide into.

a) 5 and 10   **b)** 3 and 9   **c)** 4 and 8

4 By finding a common denominator, complete these calculations.

a) $\frac{1}{3} + \frac{2}{9}$

b) $\frac{1}{4} + \frac{3}{8}$

c) $\frac{2}{5} + \frac{3}{10}$

d) $\frac{5}{6} - \frac{1}{3}$

e) $\frac{7}{12} - \frac{1}{2}$

f) $\frac{1}{6} - \frac{1}{12}$

5 Work out:

a) $\frac{1}{2} + \frac{3}{4}$

b) $\frac{3}{4} + \frac{5}{8}$

c) $\frac{11}{12} + \frac{5}{6}$

d) $1\frac{1}{2} + 1\frac{1}{4}$

Give your answers as proper fractions or mixed numbers.

6 Sarah gives $\frac{1}{4}$ of a bar of chocolate to Paul and $\frac{1}{8}$ to Rani.

a) How much chocolate did she give away?

b) How much did she keep for herself?

7 A garden has $\frac{1}{8}$ as flowerbeds and $\frac{1}{2}$ as lawn.

a) How much of the garden is this altogether?

b) How much of the garden is left?

8 Shane's string is $\frac{1}{2}$ m long, Jayne's is $\frac{1}{10}$ m and Kim's is $\frac{7}{10}$ m.
If they put them together, how far would they reach?

Investigation

9 Two fractions have a total of $1\frac{2}{5}$. What could they be?
Start by using fractions with the same denominator then go on to fractions with different denominators. You can use both addition and subtraction to help find the pairs of fractions.

Fractions of amounts

◈ Calculate fractions of quantities
◈ Multiply a fraction by a whole number

To find $\frac{1}{6}$ of £30, you divide 30 by 6.

$$£30 \div 6 = £5$$

This is the same as $\frac{1}{6} \times £30$

To find $\frac{5}{6}$ of £30:

Method 1 – Calculate $\frac{5}{6} \times \frac{30}{1}$

$\frac{5}{6}$ of £30 is the same as $\frac{5}{6} \times £30$

$$\frac{5}{6} \times \frac{30}{1} = \frac{5 \times \overset{5}{\cancel{30}}}{\underset{1}{\cancel{6}} \times 1} = \frac{25}{1} = £25$$

Method 2 – First find $\frac{1}{6}$ of £30, and then multiply by 5

$$\frac{1}{6} \text{ of } £30 = £5$$
$$\frac{5}{6} \text{ of } £30 = £5 \times 5 = £25$$

Example 1   Find $\frac{5}{8}$ of 24 hours.

$\frac{1}{8}$ of 24 = 24 ÷ 8 = 3 hours

To find $\frac{1}{8}$ we divide by 8.

$\frac{5}{8}$ of 24 = 5 × 3 hours

= 15 hours

$5 \times \frac{1}{8}$ of 24 = 5 × 3 hours

Example 2   Work out $\frac{2}{9}$ of 45 metres.

$\frac{2}{9}$ of $\frac{45}{1} = \frac{2}{9} \times \frac{45}{1}$

$\frac{2}{9} \times \frac{45}{1} = \frac{2 \times \overset{5}{\cancel{45}}}{\underset{1}{\cancel{9}} \times 1} = \frac{10}{1} = 10$ metres

Exercise 4.5

1 Work out:
   a) $\frac{1}{2}$ of 36
   b) $\frac{1}{5}$ of 35
   c) $\frac{1}{4}$ of 60 m
   d) $\frac{1}{10}$ of £120
   e) $\frac{1}{3}$ of 24 kg
   f) $\frac{1}{8}$ of 72 ml
   g) $\frac{1}{6}$ of 24 hours
   h) $\frac{1}{9}$ of £99
   i) $\frac{1}{7}$ of 35 days

Remember 'of' means the same as ×.

2 Work out:

    a) $\frac{3}{4}$ of 12   **b)** $\frac{5}{6}$ of 30

    c) $\frac{2}{7}$ of 35   **d)** $\frac{5}{8}$ of 16

3 Work out:

    a) $\frac{2}{3}$ of 21 kg   **b)** $\frac{2}{7}$ of 49 m

    c) $\frac{3}{10}$ of 60 minutes   **d)** $\frac{3}{5}$ of £55

    e) $\frac{5}{8}$ of 48 kg   **f)** $\frac{3}{4}$ of 24 mm

    g) $\frac{5}{6}$ of 60 days   **h)** $\frac{4}{9}$ of 81 mm

> Remember to include the correct unit in your answer.

4 Copy and complete these multiplication tables.

	1	2	3	5	10	20
$\times\frac{1}{2}$						
$\times\frac{1}{4}$						
$\times\frac{3}{4}$						

5 Cheru drank $\frac{3}{4}$ of a 600 ml bottle of cola. How many millilitres did she drink?

6 Sammy cut $\frac{2}{5}$ from a piece of wood 70 cm long.

    a) How much did he cut off?

    b) How much was left?

7 **a)** $\frac{1}{2}$ of an amount is £4. What is the full amount?

    b) $\frac{1}{4}$ of an amount is £6. What is the full amount?

    c) $\frac{2}{5}$ of an amount is £10. What is the full amount?

    d) $\frac{3}{4}$ of an amount is £12. What is the full amount?

Investigation

8 My answer is £24. How many different fraction calculations can you find that give this answer?

For example:

$\frac{3}{8}$ of £64 = £24 $\frac{1}{10}$ of £240 = £24

Percentages

- Know that percentage is the number of parts per 100
- Calculate simple percentages of quantities

% is short for '**per cent**'. It means 'out of one hundred'. **Percentages** are just another way of writing **hundredths**.

A percentage can be written as a decimal or a fraction.

$45\% = 0.45 = \frac{45}{100} = \frac{9}{20}$

$36\% = 0.36 = \frac{36}{100} = \frac{9}{25}$

$19\% = 0.19 = \frac{19}{100}$

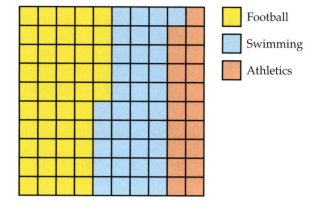

☐ Football
☐ Swimming
☐ Athletics

Here are two methods to find 21% of £300:

- **Method A**

  Find 10% of £300 = £30 | Divide by 10. |

  Find 1% of £300 = £3 | Divide by 100. |

  Use combinations of these, i.e. 21% = 10% + 10% + 1%

  So, 21% of £300 = £30 + £30 + £3 = £63

- **Method B**

  Use a calculator. Change the percentage to a decimal. Press:

  [3] [0] [0] [×] [0] [.] [2] [1] [=]

Example 1   Find 23% of 500, without using a calculator.

> 10% of 500 = 50 ——————————— | Divide by 10. |
>
> 1% of 500 = 5 —————————————— | Divide by 100. |
>
> 23% = 10% + 10% + 1% + 1% + 1%
>
>   = 50 + 50 + 5 + 5 + 5
>
>   = 115

Example 2   Using a calculator find 37% of 75

> 37% is equivalent to 0.37
>
>   [7] [5] [×] [0] [.] [3] [7] [=]   £27.75

Exercise 4.6

1 Write each of these percentages as a decimal and a fraction in its lowest terms.

a) 24% b) 15% c) 82%

d) 40% e) 41% f) 5%

2 Find:

a) 10% of £750 b) 1% of 500 kg

c) 30% of £400 d) 41% of 700 km

e) 15% of £40 f) 52% of 240 m

g) 22% of 200 mm h) 63% of 5000 seconds

3 Find:

a) 72% of 300 kg b) 54% of £62

c) 64% of 24 hours d) 83% of 1620 km

e) 18% of 50 minutes f) 47% of £350

g) 79% of 7000 h) $17\frac{1}{2}$% of £72

> Change the percentages to decimals.

4 Copy and complete this percentage table.

%	300	700	250
10%	30		
1%			
20%			
21%			
32%			

> 10% of 300 = 30

5 70% of a class of 30 pupils were football fans.

a) How many pupils were football fans?

b) How many pupils were not football fans?

6 80% of Year 8 pupils walk to school.

a) What percentage do NOT walk to school?

b) If there are 150 pupils in Year 8, how many do not walk to school?

7 Alan was paid £2500 overtime earnings last year. He has to pay 22% of this as tax. How much tax will he pay?

8 Sarah is given the choice of birthday present. She can have 15% of £1000 or 58% of £250. Which one is best?

Simplifying algebraic expressions

⊕ Know that algebra follows the same rules as arithmetic

⊕ Understand which terms you can add together and which you cannot

Key words
expression
term
like terms
simplify

When we use letters to stand in for unknown numbers, they follow the same rules as numbers.

For example: $4 + 4$ is two lots of 4 so $4 + 4 = 2 \times 4$
$n + n$ is two lots of n so $n + n = 2 \times n = 2n$

In algebra the different parts of an **expression** are called **terms**.

$4n + 3n$ are called **like terms** because they contain exactly the same letters.
Like terms can be added together. For example $4n + 3n = 7n$
$5p + 4d$ cannot be written in any simpler way. The terms $5p$ and $4d$ are not like terms.

We can **simplify** algebraic expressions by collecting like terms.

For example, $3a + 2b - b - c + a = 4a + b - c$

Example 1   If these sticks are laid end to end, what is their total length?
Simplify your answer as much as possible.

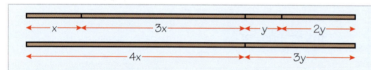

The total length is $4x + 3y$.

Example 2   Write an expression for the perimeter of each of these rectangles.

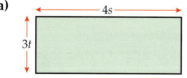

a)   Perimeter = $3t + 4s + 3t + 4s$
   = $6t + 8s$

b)   Perimeter = $t + 2s + t + 2s$
   = $2t + 4s$

c)   Perimeter = $1 + 2s + 1 + 2s$
   = $2 + 4s$

Exercise 5.1

1 If these sets of sticks are laid end to end, what is their total length?
Simplify your answers as much as possible.

a)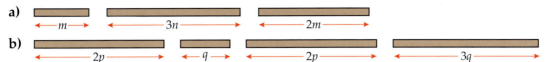

b)

2 In this addition pyramid, the expression in each brick is found by adding the two directly below it. Copy and complete the pyramid.
Remember to simplify your answers as much as possible.

a) 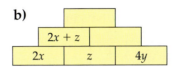   **b)**

3 Simplify the following expressions by collecting like terms.

a) $x + y + x + x + y$ **b)** $p + q - p + q + q - q + p$
c) $x + x + y - x + y$ **d)** $r + s + s + s - r + t + r + r - t$

4 Write expressions for the perimeters of these rectangles.

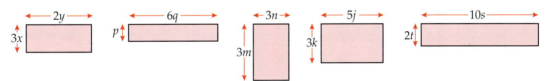

5 Write down an algebraic expression for the perimeter of:

a) the pink rectangle
b) the blue rectangle
c) the green rectangle.

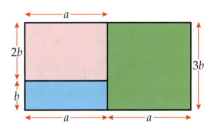

6 **a)** Write an expression for the area of each of these rectangles.
b) Find the combined area of all the rectangles. Simplify your answer as much as possible.

i) 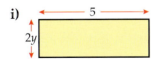   **ii)**   **iii)**

iv) 2‖ ⟵3x⟶   **v)** 2‖ ⟵3t⟶

7 Just as $2 \times 3 = 3 \times 2$, $a \times b = b \times a$.
$a \times b$ is written as ab and ab and ba are like terms.
Simplify the following expressions by collecting like terms.

a) $ab + ba + c + c$ **b)** $2b + a + ba + b$ **c)** $cd + c + ab + b + a + dc + a$

Expanding brackets

⊕ **Learn how to multiply out brackets**

Key words
simplify
expand
brackets

To **simplify** expressions it is sometimes necessary to remove (or **expand**) the **brackets**.

We can use a multiplication grid:

	7	5
2	14	10

$2(7 + 5) \rightarrow 2$

$2(7 + 5) = 14 + 10 = 24$

You can check this by doing the calculation in the brackets first:
$2(7 + 5) = 2 \times 12 = 24$.

We can expand brackets with unknowns in the same way.

For example: $3(x + 2) \rightarrow 3$

	x	2
3	$3x$	6

$= 3x + 6$

Example 1   Expand the brackets and find the value of the following:

   a) $3(5 + 9)$ **b)** $5(8 - 2)$

a) $3(5 + 9) \rightarrow 3$

	5	9
3	15	27

$3(5 + 9) \quad = 15 + 27 = 42$

b) $5(8 - 2) \rightarrow 5$

	8	2
5	40	10

$5(8 - 2) \quad = 40 - 10 = 30$

The operation in the bracket is 'subtract'.

Example 2   Expand the brackets to simplify the following expressions:

   a) $2(x + 4)$ **b)** $10(t - 5)$ **c)** $3(4 + 2x)$

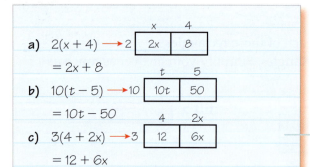

a) $2(x + 4) \rightarrow 2$

	x	4
2	$2x$	8

$= 2x + 8$

b) $10(t - 5) \rightarrow 10$

	t	5
10	$10t$	50

$= 10t - 50$

c) $3(4 + 2x) \rightarrow 3$

	4	$2x$
3	12	$6x$

$= 12 + 6x$

Remember $3 \times 2x = 3 \times 2 \times x = 6x$

Exercise 5.2

1 Expand the brackets to find the value of each of the following expressions:
   a) $2(5 + 3)$ **b)** $3(7 + 4)$ **c)** $5(9 + 10)$ **d)** $11(5 + 4)$
   e) $7(5 + 5)$ **f)** $4(8 - 5)$ **g)** $10(11 - 2)$ **h)** $8(5 - 4)$
   i) $3(6 - 6)$ **j)** $10(15 - 4)$ **k)** $12(3 + 10)$

2 Check your answers to **Q1** by doing the calculation in the brackets first.
For example: $2(5 + 3) = 2 \times 8 = 16$.

3 Simplify the following algebraic expressions:

> Remember to subtract.

a) $2(t + 7)$ b) $3(x - 1)$

c) $10(p + 5)$ d) $8(t - 2)$ e) $4(x + 5)$ f) $3(x - 20)$

g) $5(10 + m)$ h) $5(6 - m)$ i) $12(10 + b)$ j) $7(4 - p)$

4 For each rectangle:
Write an expression for the area, using brackets.
Expand the brackets to simplify your expression.

> Area of a rectangle = length × width

2x + 4, 3

3b + 2

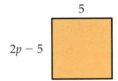
10

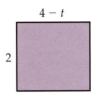
4 − t, 2

y − 3, 4

2p − 5, 5

5 For each rectangle in **Q4**:
– Write an expression for the perimeter
– Simplify your expression.

6 Each algebraic expression in set A matches to an equivalent expression in set B.
Write down the matching pairs.

Set A

| $4(x + y)$ | $3(2 - r)$ | $4(s - 2t)$ |

| $3(m + 2)$ | $2(3y + 3)$ | $4(x - y)$ |

Set B

| $6y + 6$ | $4x - 4y$ | $4s - 8t$ |

| $3m + 6$ | $6 - 3r$ | $4x + 4y$ |

7 To work out the amount of tax you must pay you use the formula:
Tax $= 0.22(P - 4000)$
where P is the money you earn.
Expand the bracket to simplify the formula.

8 Expand the brackets and simplify the following expressions:

a) $4(5 + x + 2)$ b) $3(12 + 2p - 4)$ c) $7(2y + 3 - y)$

d) $5(2m - 4 + m)$ e) $10(9 + 3r - 3 + 2r)$ f) $6(4 - 2s + 6 + 3s)$

Check your answers by simplifying the expression in the brackets first.

Substitution

◈ Substitute values into algebraic expressions

◈ Recognise that algebra follows the same rules as arithmetic

---

We use letters in algebra to represent numbers.

Substitution is replacing the letters with numbers.

Look at the **expression** $2ab$.

If $a = 3$ and $b = 10$ we can calculate the value of the expression $2ab$ by replacing the letters with their values.

> Remember $2ab$ means $2 \times a \times b$

$$2ab \quad = 2 \times a \times b$$

$$= 2 \times 3 \times 10$$

$$= 60$$

$$= 60 \text{ square units}$$

Total area $= 2ab$

Example 1 Find the value of the following if $a = 7$, $b = 10$ and $c = 2$

a) ab **b)** $2a - 3c$ **c)** $\dfrac{b}{c}$

a) $ab = 7 \times 10$

$= 70$

> Remember ab means $a \times b$

b) $2a - 3c = 2 \times 7 - 3 \times 2$

$= 14 - 6$

$= 8$

> Remember the order of operations: do the multiplication first.

c) $\dfrac{b}{c} = \dfrac{10}{2}$

$= 5$

> Remember $\dfrac{b}{c}$ means $b \div c$

Example 2 Simplify the following expressions and then find the value when $x = 20$ and $y = 5$.

a) $2x + 3y - x$ **b)** $3 + 2(x + y)$

a) $2x + 3y - x$

$= x + 3y$

$= 20 + 3 \times 5$

$= 35$

> Collect like terms.

> Substitute the values of x and y.

b) $3 + 2(x + y)$

$= 3 + 2x + 2y$

$= 3 + 2 \times 20 + 2 \times 5$

$= 3 + 40 + 10$

$= 53$

> Expand the brackets.

> Substitute the values of x and y.

Exercise 5.3

1 Find the value of the following if $a = 5$, $b = 7$ and $c = 9$.

a) $a + 4$ b) $4c$ c) $9 - a$ d) $\dfrac{c}{3}$

e) $a + c$ f) $c - a$ g) ab h) ac

2 Simplify each of the following expressions.

a) $2x - x + 2y$ b) $3xy + 7yx$ —————— Remember xy is the same as yx.

c) $2x - 3y + 10x$ d) $2 + 5(x - 4)$

e) $4x + 4y - 4x$ f) $3(5x + 2y - 3x)$

3 Find the value of each of the expressions in **Q2** when: $x = 10$ and $y = 7$.

4 In a recipe for making a cake the amount of flour used (in grams) is: $75e + 30$ where e is the number of eggs used. Find the amount of flour required when:

a) 3 eggs b) 10 eggs c) 15 eggs are used.

5 At a Rugby tournament 2 points are awarded for a try and 1 point for a goal.
The number of points for each team is calculated using the expression: $2w + d$
where w = number of tries and d = number of goals.
The table shows the number of tries and goals for each team at the tournament:

Team	Number of tries (w)	Number of goals (d)
Sherborne Primary School	7	3
Carling College	2	5
Iris Institution	1	8
Pould Park	3	3

a) Work out the number of points for each team.

b) Who won the tournament?

6 For a mobile phone bill the cost in pounds per month is:
$0.1T + 0.2C + 10$ where T is the number of text messages sent and C is the total minutes spent on calls.
Find the cost of the bill for each of these people:

	Name	Number of texts sent (T)	Minutes spent on calls (C)
a)	Ameer	12	20
b)	Ben	20	30
c)	Carina	45	5
d)	Drew	2	35
e)	Emily	30	30

Investigation

7 Find values of a, b and c that satisfy the following equation: $2a + 3b - c = 26$

a, b and c are all integers (whole numbers).

For example:

$2 \times 4 + 3 \times 10 - 12 = 26$

$a = 4, b = 10, c = 12$

There are many possible combinations of values. Try fixing one value and finding values for the other two.

Solving equations

◈ Solve simple equations

Key words
equation
equal
solve
inverse operations
balance

An **equation** is when two things are **equal** . An equals sign means that both sides are the same.

For example $x + 3 = 12$ is an equation.

One way to **solve** equations is to use **inverse operations** as we did in Year 7. Another way is to **balance** the equation by doing the same thing to both sides.

Example   Solve the equation $x + 3 = 12$.

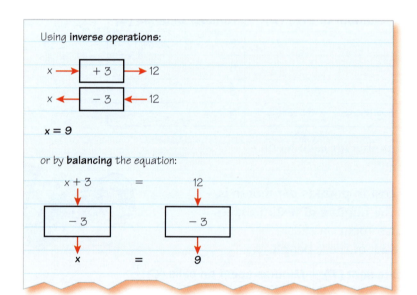

To remove $+3$ we must subtract 3.

Exercise 5.4

1 Solve each of the following equations:

a) $a + 5 = 7$ b) $b + 12 = 13$ c) $c + 9 = 12$ d) $d + 3 = 20$

e) $e + 7 = 12$ f) $x - 3 = 7$ g) $q - 4 = 12$ h) $m - 10 = 8$

2 Solve each of the following equations:

a) $4 = x + 1$ b) $7 = x + 5$ c) $18 = q + 12$

d) $7 = r - 4$ e) $12 = v - 8$ f) $25 = f - 50$

g) $18 = m + 12$ h) $100 = h - 5$ i) $25 = p + 25$

It doesn't matter which side of the equals sign the unknown is on.

3 To check that you have solved the equation correctly, substitute your value for x back into the original equation.

For example, in the Example we solved $x + 3 = 12$ and found that $x = 9$.

To check this:

    Left-hand Side $= x + 3$
                    $= 9 + 3$ —————— Substitute $x = 9$ into the equation.
                    $= 12$
                    $=$ Right-hand Side

Check your answers to **Q2** in this way.

4 Lord Number collects miniature china vegetables! He keeps them all locked up in his safe. The combination on his safe each day is the solution to an equation. He forms the equations by writing: $x -$ month number $=$ date.

For example on the 5th of July which is the 5th day of the 7th month the equation he uses is: $x - 7 = 5$.

  a) What is his combination on the 5th July?

  b) What is his combination on the 20th January?

  c) List all the different combinations he uses in January. (You should spot a pattern!)

  d) Investigate the combinations for the other months.

  e) Which combinations does he use only once?

  f) What dates are these for?

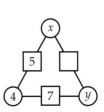

5 In this addition pyramid, the value in each brick is found by adding the two bricks directly below it.

	14	
$2n + 2$	3	$1 - n$

  a) Copy the addition pyramid and fill in the expressions in the empty bricks.

  b) Add your two expressions together and simplify.

  c) From the addition pyramid, what is your expression from part **b)** equal to? Write this as an equation.

  d) Solve your equation to find n.

6 In this arithmagon the number in each square is the sum of the two numbers in the circles on either side.

  a) Write an equation for x. $x + 4 = ?$
      Solve your equation to find x.

  b) Write an equation for y.
      Solve your equation to find y.

  c) What value should be written in the empty square?

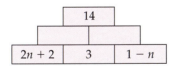

More solving equations

⊕ Solve equations involving multiples of *x*

We can **solve equations** such as $3x = 21$ or $2a + 5 = 13$ by using inverse operations or the balancing method.

Example 1   Solve the equation $2a + 5 = 13$

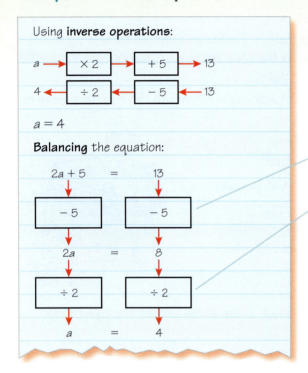

The inverse of 'add 5' is 'subtract 5'.

Divide both sides by 2.

Check your answer by substituting $a = 4$ into the equation.
Left-hand Side $= 2a + 5$
$= 2 \times 4 + 5$
$= 8 + 5$
$= 13$
$=$ Right-hand Side

Example 2   Three books of the same price cost £21.
   a)   Write out an equation using *x* to stand in for the cost of one book.
   b)   Solve the equation to find the cost of one book.

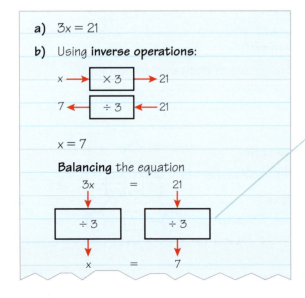

Divide both sides by 3.

Check your answer by substituting $x = 7$ into the equation.
Left-hand Side $= 3x$
$= 3 \times 7$
$= 21$
$=$ Right-hand Side

Exercise 5.5

1 Solve the following equations. Check your answer by substituting back into the original equation.

a) $4a = 12$ b) $7b = 21$ c) $12c = 124$

d) $9d = 27$ e) $10e = 40$ f) $8f = 32$

g) $24 = 6g$ h) $77 = 11h$ i) $35 = 7i$

2 Find the value of x in each of the following equations. Check your answers by substituting your value of x back into the original equation.

a) $2x + 3 = 11$ b) $3x + 5 = 14$ c) $5x + 9 = 19$

d) $7x + 6 = 27$ e) $12x + 5 = 29$ f) $2y - 2 = 18$

g) $3y - 5 = 25$ h) $6y - 35 = 1$ i) $10y - 20 = 30$

3 10 chocolate bars weigh 500 g.

a) Write an equation using w to stand in for the mass (in grams) of a chocolate bar.

b) Solve the equation to find the mass of each chocolate bar.

4 Lady Diamond keeps her precious jewels in a safe.
The combination on her safe each day is the solution to an equation.

She forms the equation by writing:

$$2x - \text{month number} = \text{date}$$

For example on the 5th July which is the 5th day of the 7th month the equation she uses is:

$$2x - 7 = 5$$

a) What is the combination on the 5th July?

b) What is the combination on the 21st January?

c) List all the different combinations she uses in January. Can you spot a pattern?

d) Investigate the combinations for other months.

e) Look at your answers to **Q4** in Exercise 5.4. How are Lady Diamond and Lord Number's combinations connected?

f) Explain why you think this is.

5 Two coffees and one tea cost £3.50. One tea costs £1.

a) Write an equation for this information, using x to stand in for the cost of one coffee.

b) Solve your equation to find the cost of one coffee.

> *Cosy Café*
> 2 coffees
> 1 tea
> £3.50

6 In this addition pyramid, the number in each brick is found by adding the two bricks directly below it.

a) Copy and complete this addition pyramid.

		55		
	30			
15			10	

b) Copy this addition pyramid and write the correct expressions in the empty bricks.

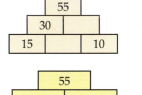

		55		
$4n + 1$		5		$6n + 4$

c) Add your two expressions together and simplify them.

d) From the addition pyramid, what is your expression from part **c)** equal to? Write this as an equation.

e) Solve your equation to find n.

Formulae

◈ Use formulae from mathematics and other subjects

A **formula** is a way of expressing a relationship.

For example the formula:

Area of a rectangle = length × width

expresses a relationship between the area, length and width of a rectangle.

We can **substitute** values into formulae to find the value of an **unknown** .

Example 1   The formula for finding the area of a rectangle is:

Area of a rectangle = length × width

Find the area of a rectangle with length 11 cm and width 12 cm.

$A = 11 \times 12$ —————— Substitute the values 11 and 12 into the formula.

$= 132 \, cm^2$

Both lengths are in cm,
so the area is in cm^2.

Example 2   Write a formula for the number of days in a given number of weeks.

There are 7 days in a week.

So the formula is:

Number of days = 7 × number of weeks.

Exercise 5.6

❶ Find the area of these rectangles by substituting the values for length and width into the formula: Area of a rectangle = length × width

  a) length 5 cm, width 10 cm

  b) length 4 cm, width 8 cm

  c) length 7 m, width 5 m

  d) length 2.5 cm, width 8 cm.

❷ **a)** Write a formula for the number of pence in a given number of pounds.

  b) Use your formula to find the number of pence in:

   i) £5   **ii)** £10   **iii)** £7.50

3 **a)** Write a formula for a person's age in months if you know their age in years.

   b) Use your formula to calculate the age in months of a person who is exactly:
   i) 5 years old   **ii)** 10.5 years old   **iii)** 12 years old.

4 In a café, a glass of orange juice costs £1.20.

   a) Write a formula that a waiter could use to calculate the cost of a number of glasses of orange juice.

   b) One group of people orders 7 glasses of orange juice. Use your formula to calculate the cost.

5 A bank changes pounds to euros using this formula:

   number of euros = 1.5 × number of pounds

Copy and complete the table for pounds converted to euros.

Pounds (£)	Euros (€)
100	
75	
150	
180	
	300
	600

6 A building surveyor changes measurements in cm to measurements in m using this formula: Measurement in m = measurement in cm ÷ 100

   a) Use the formula to convert these measurements to metres:
   i) 1000 cm   **ii)** 2500 cm   **iii)** 8600 cm

   b) Explain why this formula works.

   c) Write a formula to convert m to cm. Check your formula works by converting your answers to part **a)** back to cm.

Investigation

7 **a)** Write a formula for the perimeter of this rectangle.
   b) Simplify your formula as much as possible.
   c) Use your formula to find the perimeters of these rectangles.

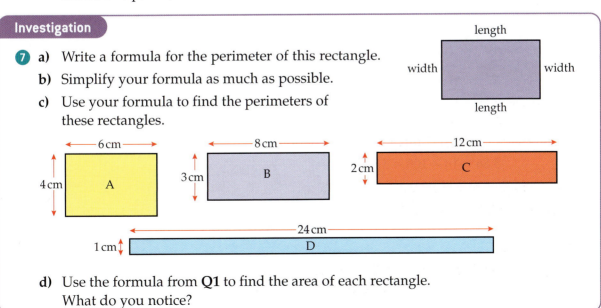

   d) Use the formula from **Q1** to find the area of each rectangle. What do you notice?

Multiplying and dividing by 10, 100, 1000

Key words
digit
place value grid
decimal point

◈ Know how to multiply whole numbers and decimals by 10, 100, 1000
◈ Know how to divide whole numbers and decimals by 10, 100, 1000

When multiplying a number by 10, 100 or 1000, the **digits** slide 1, 2 or 3 places to the left on a **place value grid** .

TTh	Th	H	T	U	.	t	h

17.6 → 1 7 . 6

17.6 × 10 = → 1 7 6

17.6 × 100 or 176 × 10 = → 1 7 6 0

17.6 × 1000 or 176 × 100 or 1760 × 10 = → 1 7 6 0 0

176 is the same as 176.0, but the decimal point is usually not shown.

Remember to put zeros in the units or tens column if necessary. You don't need to include the **decimal point** after 176, 1760 and 17 600.

When dividing a number by 10, 100 or 1000, the digits slide 1, 2 or 3 places to the right on a place value grid.

Th	H	T	U	.	t	h

4600 → 4 6 0 0

4600 ÷ 10 = → 4 6 0

4600 ÷ 100 or 460 ÷ 10 = → 4 6

4600 ÷ 1000 or 460 ÷ 100 or 46 ÷ 10 = → 4 . 6

460 ÷ 1000 or 46 ÷ 100 or 4.6 ÷ 10 = → 0 . 4 6

4600 is the same as 4600.0, but the decimal point is usually not shown.

Multiplying by 10, and then by 10 again is equivalent to multiplying by 100.
Dividing by 100, and then by 10 is equivalent to dividing by 1000.

Example 1 Work out:

a) 6.7×100 b) $13.2 \div 10$

a) $6.7 \times 100 = 670$ —— The digits move two places to the left, so we need to put a '0' in the units column.

b) $13.2 \div 10 = 1.32$ —— We move the digits one place to the right.

Example 2 Copy and complete this calculation.
$3.86 \times 10 \times 100 = 3.86 \times \square = \square$

$3.86 \times 10 \times 100 = 3.86 \times \mathbf{1000} = \mathbf{3860}$

$\times 10 \times 100 = \times 1000$

Exercise 6.1

1 Copy and complete this place value grid.

Th	H	T	U	.	t	h	th
	3	6	.	2	9		

$36.29 \times 10 =$

$36.29 \times 100 =$

Th	H	T	U	.	t	h	th
		3	6	.	2	9	

$36.29 \div 10 =$

$36.29 \div 100 =$

2 Make a place value grid for 6.7 and use it to answer the following.

a) 6.7×10 b) $6.7 \div 10$ c) $6.7 \div 100$

d) 6.7×100 e) 6.7×1000 f) $6.7 \div 10 \div 10$

3 Complete the following multiplications.

a) 3.24×10 b) 6.3×100 You can use a place value grid to help.

c) 0.59×1000 d) 15.2×100 e) 0.8×10

4 Complete the following divisions.

a) $4.2 \div 10$ b) $21.5 \div 10$ You can use a place value grid to help.

c) $30.3 \div 100$ d) $6.42 \div 100$ e) $3 \div 10$

5 Complete the following calculations.

a) $36.2 \div 10$ b) 7×1000 c) 4.9×100

d) $2.13 \div 10$ e) $8 \div 10$ f) $14.6 \div 100$

g) 0.32×10 h) 4.6×100 i) $213 \div 1000$

j) 0.03×10

6 Find the missing value in each of the following:

a) $4.93 \times \square = 49.3$ b) $6.21 \div \square = 0.621$ c) $7.3 \div \square = 0.073$

d) $5.8 \times \square = 580$ e) $3.2 \times \square = 320$ f) $\square \times 0.24 = 2.4$

g) $100 \times \square = 458$ h) $\square \div 10 = 0.2$

7 Copy and complete:

a) $3.24 \times 10 \times 100 = 3.24 \times \square$

b) $15.3 \div 10 \div 10 = 15.3 \div \square$

c) Now work out the answers to parts **a)** and **b)** above.

8 In a sweet shop a box of 10 chocolate bars cost £3.70. How much does each bar cost?

9 A hundred people are going to Alton Water Park.

a) It cost £3.25 each for the bus. How much is that altogether?

b) They pay a total of £1650 to get into the park. How much is that each?

10 If 1 sweet has a mass of 3 g, what will be the mass of:

a) 10 sweets   b) 100 sweets   c) 1000 sweets?

Now work out the mass of:

d) 200 sweets   e) 500 sweets.

Scales and measures

◈ Read and interpret scales on a range of measuring instruments

◈ Solve everyday problems involving length, mass and time

To read a **scale** accurately:

• work out the value between the main **divisions** on the scale

• work out the value between the small divisions (or **subdivisions**) on the scale.

The distance between main divisions is 5. There are 5 subdivisions. The distance between subdivisions is $5 \div 5 = 1$.

The weight is 2.6 kg

The distance between main divisions is 1. There are 10 subdivisions. The distance between subdivisions is $1 \div 10 = 0.1$.

The temperature reading is 18°

Example Write the position of the pointers on each measuring instrument.

a)

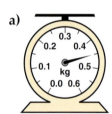

b)

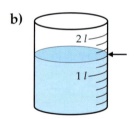

c)

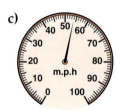

a) The distance between main divisions is 0.1. There are 2 subdivisions. The distance between subdivisions is $0.1 \div 2 = 0.05$. The pointer shows 0.45 kg.

b) The distance between main divisions is 1. There are 5 subdivisions. The distance between subdivisions is $1 \div 5 = 0.2$. The pointer shows 1.4 ℓ.

c) The distance between main divisions is 10. There are 10 subdivisions. The distance between subdivisions is $10 \div 10 = 1$. The pointer shows 54 m.p.h.

Exercise 6.2

1 Write the position of the pointers on each of these lines.

a)

```
|++++++++++++++++++++++++++++|
0    ↑    ↑       ↑   ↑      ↑    100
     A    B       C   D      E
```

b)

```
|++++++++++++++++++++++++++++++++++++++|
100  ↑   ↑    ↑    ↑          ↑    200
     A   B    C    D          E
```

c)

```
|++++++++++++++++++++++++++++++++++++++++|
30    ↑      ↑        ↑     ↑   ↑ 40
      A      B        C     D   E
```

2 Write the position of the pointers on each measuring instrument.

a)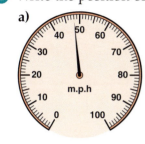

b)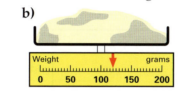

c)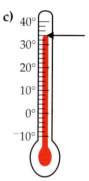

d)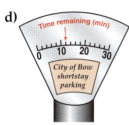

e)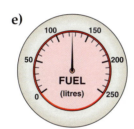

f)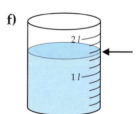

3 Write the position of the pointers if the
ends of the line are marked:

a) 0 and 100

b) 0 and 1000

c) 0 and 20

d) 3 and 4

e) 50 and 60

f) ⁻5 and ⁺5

```
   D      C        B       A      E
   ↓      ↓        ↓       ↓      ↓
|++++++++++++++++++++++++++++++++++++++++|
```

4 Passengers must book in 90 minutes before their flight is due for take off.

a) What is the latest time to book in for a flight at 1155?

b) What is the latest time to book in for a flight at 1915?

5 A machine bends wire into paperclips.
It uses 960 cm of wire to make 120 paperclips.
How much wire does it use for each clip?

6 A box of 100 fruit drops weighs 750 grams.
The box weighs 100 grams.
How much does each fruit drop weigh?

> First find the weight of **just** the fruit drops by subtracting the weight of the box (100 g) from the total weight of the box of fruit drops (750 g).

Area of rectangles

◈ Know and use the formula for the area of a rectangle

Key words
area
square millimetre (mm²)
square centimetre (cm²)
square metre (m²)
square kilometre (km²)

The **area** is the space inside a two dimensional shape.

We use **squares** to measure area because they don't leave any gaps.

A **square millimetre** (mm²) is about this size:

A **square centimetre** (cm²) is about this size:

Other units of area are the **square metre** (m²) and **square kilometre** (km²).

We can calculate the area of a rectangle by multiplying the length by the width.

Area = length × width

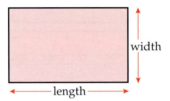

Example   Find the area of this shape:

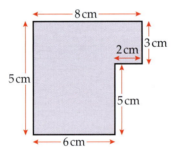

This can be done in different ways. Either break the shape down into rectangles or take the missing rectangle out of the corner of the large rectangle.

Breaking the shape into two rectangles:

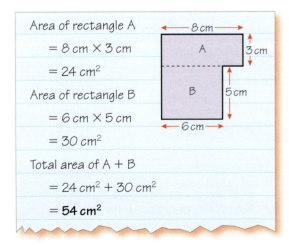

Area of rectangle A

   = 8 cm × 3 cm

   = 24 cm²

Area of rectangle B

   = 6 cm × 5 cm

   = 30 cm²

Total area of A + B

   = 24 cm² + 30 cm²

   = **54 cm²**

Take a rectangle away from the large rectangle:

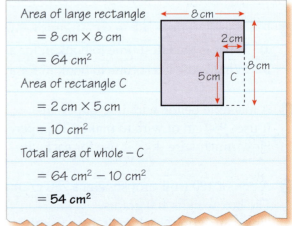

Area of large rectangle

   = 8 cm × 8 cm

   = 64 cm²

Area of rectangle C

   = 2 cm × 5 cm

   = 10 cm²

Total area of whole − C

   = 64 cm² − 10 cm²

   = **54 cm²**

Exercise 6.3

1 Calculate or count squares to find the areas of these rectangles.
Make sure you record the units too.

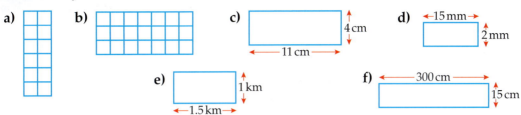

2 Find the area of each of these rectangles.

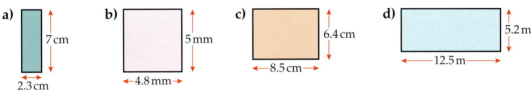

3 Find the area of each of these shapes.

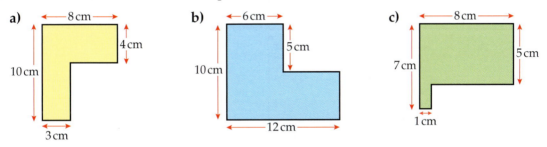

4 Work out the missing dimension in each of these rectangles. The areas are given inside each rectangle.

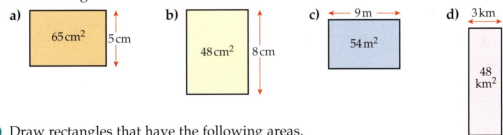

5 Draw rectangles that have the following areas.
Choose lengths and widths that are whole numbers greater than 1.
Show clearly on your drawings the dimensions of your rectangles.

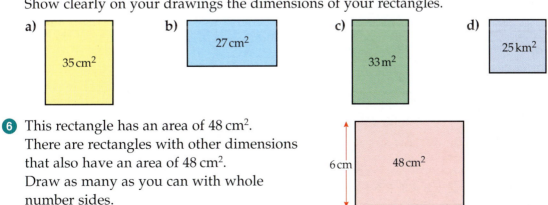

6 This rectangle has an area of 48 cm².
There are rectangles with other dimensions that also have an area of 48 cm².
Draw as many as you can with whole number sides.

6 cm — 48 cm² — 8 cm

Perimeter and area

⊕ Calculate the perimeter and area of shapes made from rectangles

---

The **perimeter** is the distance all the way around the edge of a shape.

The perimeter of a rectangle is 2 × length + 2 × width.

The **area** is the space inside a two dimensional shape.
The area of a rectangle is length × width.

This rectangle has a perimeter of $2 \times 3 + 2 \times 1 = 8$ cm
It has an area of $3 \times 1 = 3$ cm²

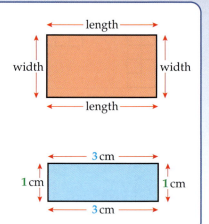

Example   Work out the area and perimeter of this shape.

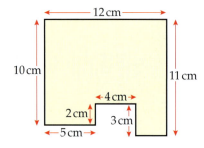

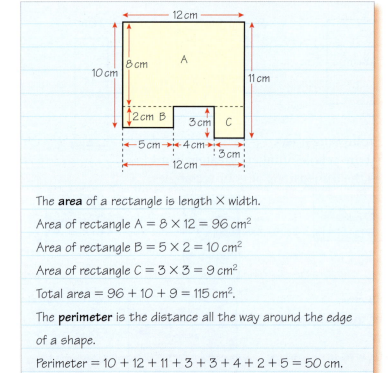

The **area** of a rectangle is length × width.

Area of rectangle A = 8 × 12 = 96 cm²

Area of rectangle B = 5 × 2 = 10 cm²

Area of rectangle C = 3 × 3 = 9 cm²

Total area = 96 + 10 + 9 = 115 cm².

The **perimeter** is the distance all the way around the edge of a shape.

Perimeter = 10 + 12 + 11 + 3 + 3 + 4 + 2 + 5 = 50 cm.

First, split the shape into rectangles and find the size of the missing measurements.

Remember to add in order around the shape so that no measurements are missed.

Exercise 6.4

1 Calculate the perimeters of these shapes.

a)
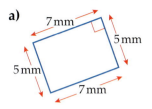

b)
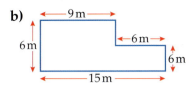

c)
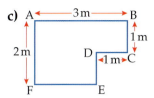

2 Find the area and perimeter of each of these shapes.

a)
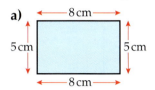

b)
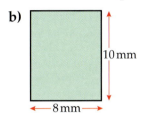

c)
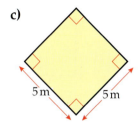

3 Measure these rectangles and work out the perimeter and the area of each one.

a)
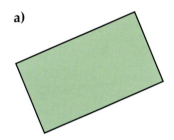

b)
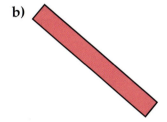

c)
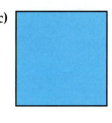

4 Draw a rectangle with a perimeter of 16 cm. Work out the area of your rectangle.

5 Play with a partner. You will need one normal dice, and a piece of squared paper each. In one game, you have four turns each. Take turns to throw the dice and shade in that number of squares on your piece of paper. The squares must be joined to each other by at least one side. At the end of the game the winner is the one whose shape has the largest perimeter.

6 Find the width and length of rectangles that have:

> Use squared paper to help you.

    a) area of 12 cm² and perimeter of 14 cm   **b)** area of 20 cm² and perimeter of 18 cm.

Investigation

7 Using squared paper, draw rectangles with different whole number dimensions that have an area of 36 square units.

Work out the perimeter of each of your rectangles.

Which rectangle has the largest perimeter? Which rectangle has the smallest perimeter?

Now choose your own area. Investigate the largest and smallest perimeters for rectangles with this area.

Write a hint for someone who is trying to make a rectangle from 100 squares with the smallest possible perimeter.

6.5 Area of triangles

⊕ Understand and use the formula for finding the area of any triangle

Key words
area
formula
base
height
perpendicular

We can calculate the **area** of a rectangle by multiplying the length by the width. The **formula** is Area = length × width

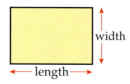

Shapes that do not have right angles have slant heights as well as **perpendicular** heights. We need to know the perpendicular height of the shape to find the area.

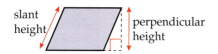

A rectangle has length and width but a triangle has **base** and **height**.
We can calculate the area of a right-angled triangle by imagining it is half of a rectangle. The formula is Area = $\frac{1}{2}$ (base × height)

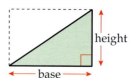

We can calculate the area of a non-right-angled triangle by splitting it into two right-angled triangles, or by imagining it is half a rectangle.
The formula is :
Area = $\frac{1}{2}$ (base × perpendicular height)

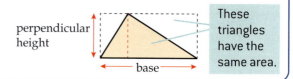

These triangles have the same area.

Example 1 Calculate the area of this triangle.

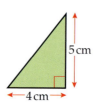

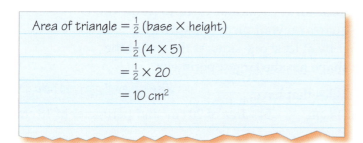

Area of triangle = $\frac{1}{2}$ (base × height)

$= \frac{1}{2}$ (4 × 5)

$= \frac{1}{2}$ × 20

$= 10$ cm²

Example 2 Calculate the area of this triangle.

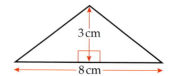

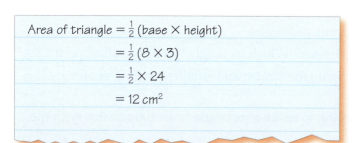

Area of triangle = $\frac{1}{2}$ (base × height)

$= \frac{1}{2}$ (8 × 3)

$= \frac{1}{2}$ × 24

$= 12$ cm²

Exercise 6.5

1 Calculate the area of each triangle.

a)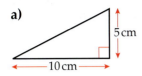

b)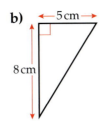

c)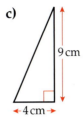

2 Calculate the area of each triangle. Their perpendicular heights are labelled.

a)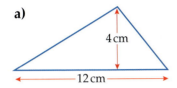

b)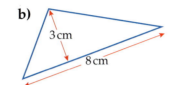

c)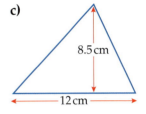

3 Calculate the area of each triangle. You will first need to measure the base and height of each triangle in centimetres.

a)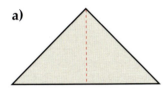

b)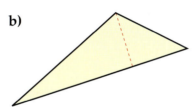

4 Draw a triangle of your own with a whole number base and a whole number height. Give it to your partner and ask them to calculate its area.

5 This boat has a triangular sail with base 2 m and height 4 m. Find the area of the sail.

6 a) Make a sketch of this rhombus and draw in the diagonals

b) What shapes have been formed by the diagonals?

c) How can you use these shapes to work out the area of the rhombus?

d) Find the area of the rhombus.

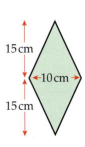

6.6 Surface area

Key words
surface area
net

- ◈ Calculate surface areas of cuboids
- ◈ Use surface area to solve problems

A cube and a cuboid have six faces.

The total area of the faces is called the **surface area** .

Think of a dice: the faces are numbered from 1 to 6.

Drawing the **net** of a cuboid helps us to make sure that we have added up the areas of all the faces.

You can see that there are two blue rectangles, two purple rectangles and two orange rectangles.

The general formula for the surface area of a cuboid is

2(length × width) + 2(length × height) + 2(width × height)
or $2lw + 2lh + 2wh$.

Example   Calculate the surface area of this cuboid.

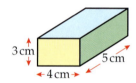

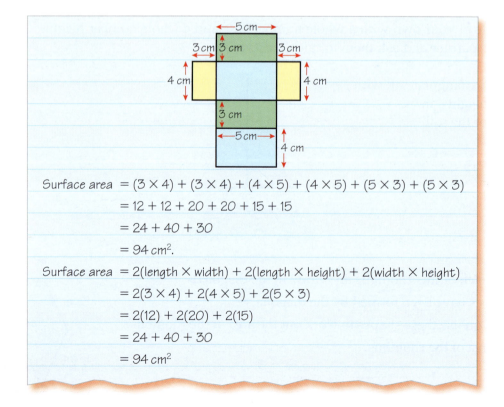

Draw the net for the cuboid and find the area of each of the faces.

Surface area = (3 × 4) + (3 × 4) + (4 × 5) + (4 × 5) + (5 × 3) + (5 × 3)

   = 12 + 12 + 20 + 20 + 15 + 15

   = 24 + 40 + 30

   = 94 cm².

Surface area = 2(length × width) + 2(length × height) + 2(width × height)

   = 2(3 × 4) + 2(4 × 5) + 2(5 × 3)

   = 2(12) + 2(20) + 2(15)

   = 24 + 40 + 30

   = 94 cm²

Alternatively, substitute the values for the length, width and height into the formula.

Exercise 6.6

1 Make the following cuboids with linking cubes. Work out their surface areas.

a)
b)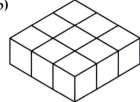
c)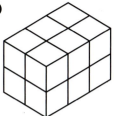

> The surface area of a shape is the area of **all** its faces.

2 Here are the nets of different cuboids. For each one, find the area of the yellow, blue and green shapes. Then find the total surface area for each net.

a)
3 mm ← → ← 10 mm → ← → 3 mm
5 mm
5 mm
3 mm
5 mm
3 mm

b)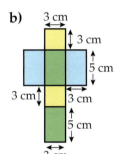
3 cm
3 cm
5 cm
3 cm
3 cm
5 cm
5 cm
3 cm

c)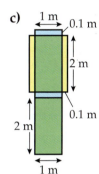
1 m
0.1 m
2 m
0.1 m
2 m
1 m

d)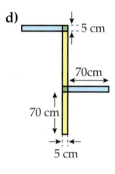
5 cm
70cm
70 cm
5 cm

3 Calculate the surface area of each of these shapes.

> You might find it helpful to draw the nets.

a)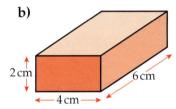
4 cm

b)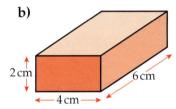
2 cm
4 cm
6 cm

c)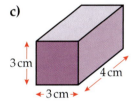
3 cm
4 cm
3 cm

d)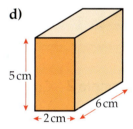
5 cm
2 cm
6 cm

4 The diagram shows a cuboid made from 24 cubes. It has a surface area of 70 square units.

a) Using 24 linking cubes, make a different cuboid and draw it on isometric paper. Work out its surface area.

b) Investigate different cuboids made with the 24 cubes. Draw your cuboids on isometric paper and write down the surface area each time.

c) Which cuboid has the biggest surface area? Which cuboid has the smallest surface area?

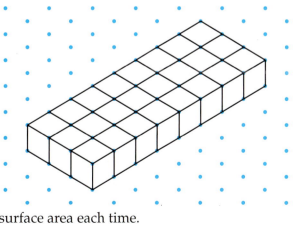

Mappings

◈ Draw simple mapping diagrams

A **mapping** is a rule that connects one set of numbers to another set of numbers.

We write a mapping like this: $x \rightarrow x + 5$

We can show a mapping using a **mapping diagram** :

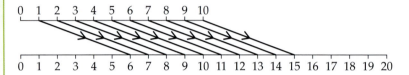

Each number on the top number line maps to the number it becomes when we add 5.

We can show the same mapping in a list like this:

$0 \rightarrow 5$	0 maps to 5
$1 \rightarrow 6$	1 maps to 6
$2 \rightarrow 7$	2 maps to 7
$3 \rightarrow 8$	3 maps to 8

Each number on the left maps to the number on the right by adding 5.

Example 1 On a pair of number lines from 0 to 18 draw the mapping:

$x \rightarrow x - 3$

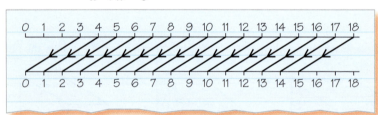

To complete the mapping diagram we need to show what each number on the top number line will become when we subtract 3.

Example 2 Copy and complete the mapping list below for: $x \rightarrow 11x$

$1 \rightarrow 11$ $2 \rightarrow ...$ $3 \rightarrow ....$ $4 \rightarrow .....$

Remember $11x$ means $11 \times x$

$1 \rightarrow 11$

$2 \rightarrow 22$ ————————

$3 \rightarrow 33$

$4 \rightarrow 44$

To work out what 2 maps to we calculate:
$11 \times 2 = 22$

Exercise 7.1

1 Copy and complete a separate mapping diagram to show each of the following mappings.

a) $x \rightarrow x + 2$ **b)** $x \rightarrow x - 1$ **c)** $x \rightarrow x + 4$ **d)** $x \rightarrow x - 3$

0 1 2 3 4 5 6 7 8 9 10

0 1 2 3 4 5 6 7 8 9 10

2 Draw a mapping diagram to show each of the following mappings. Draw each pair of number lines from 1 to 10.

a) $x \rightarrow 2x$ b) $x \rightarrow 3x$ c) $x \rightarrow 4x$

3 Copy and complete the mapping list for each of the following:

a) $x \rightarrow x + 7$
$1 \rightarrow \ldots$
$2 \rightarrow \ldots$
$3 \rightarrow \ldots$
$4 \rightarrow \ldots$

b) $x \rightarrow x + 8$
$1 \rightarrow \ldots$
$2 \rightarrow \ldots$
$3 \rightarrow \ldots$
$4 \rightarrow \ldots$

c) $x \rightarrow 3x$
$1 \rightarrow \ldots$
$2 \rightarrow \ldots$
$3 \rightarrow \ldots$
$4 \rightarrow \ldots$

d) $x \rightarrow 5x$
$1 \rightarrow \ldots$
$2 \rightarrow \ldots$
$3 \rightarrow \ldots$
$4 \rightarrow \ldots$

4 Copy and complete these mapping lists:

a) $x \rightarrow x - 7$
$10 \rightarrow \ldots$
$11 \rightarrow \ldots$
$12 \rightarrow \ldots$
$13 \rightarrow \ldots$

b) $x \rightarrow x - 8$
$10 \rightarrow \ldots$
$11 \rightarrow \ldots$
$12 \rightarrow \ldots$
$13 \rightarrow \ldots$

c) $x \rightarrow x \div 5$
$5 \rightarrow \ldots$
$10 \rightarrow \ldots$
$15 \rightarrow \ldots$
$20 \rightarrow \ldots$

d) $x \rightarrow x \div 3$
$3 \rightarrow \ldots$
$6 \rightarrow \ldots$
$9 \rightarrow \ldots$
$12 \rightarrow \ldots$

5 A car manual gives the following table and mapping to show how far you can drive per litre of petrol.

a) If you had 5 litres of petrol how far could you drive?

b) If you had 40 litres of petrol how far could you drive?

c) If you had 7 litres of petrol how far could you drive?

Litres		No. of miles
1	→	7
10	→	70
15	→	105
20	→	140
50	→	350

Mapping $x \rightarrow 7x$

6 A phone box gives the following table and mapping to show how many minutes of calls you can make for different amounts of money.

a) If you had 60p how many minutes of calls could you make?

b) If you had 80p how many minutes of calls could you make?

Pence		No. of minutes
10	→	1
20	→	2
30	→	3
50	→	5

Mapping $x \rightarrow \dfrac{x}{10}$

7 If you mapped a positive number using the mappings below would you end up with a smaller or larger number?

a) $x \rightarrow x + 5$ b) $x \rightarrow 3x$

c) $x \rightarrow x \div 10$ d) $x \rightarrow x - 4$

e) $x \rightarrow x \div 3$ f) $x \rightarrow x + 12$

g) $x \rightarrow x - 6$ h) $x \rightarrow 8x$

Describe any patterns that you notice.

> You may wish to try each of the mappings out on a particular number to help you decide.

Identifying mappings

⊕ Identify a mapping from a diagram

A **mapping diagram** tells us about the relationship between two sets of numbers.

We can identify the **mapping** shown in a mapping diagram by looking at the relationship between pairs of numbers.

For example:

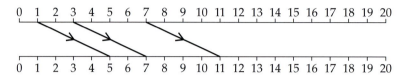

In this diagram we can see that:

$1 \rightarrow 5$

$3 \rightarrow 7$

$7 \rightarrow 11$

The rule to get from the left-hand column of numbers to the right-hand column is 'add 4'.

So the mapping is $x \rightarrow x + 4$

Example 1   Identify the mapping shown in the mapping diagram.

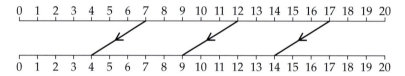

The mapping is: $x \rightarrow x - 3$

By looking at pairs of numbers we can see that 3 is subtracted from each number. The rule is 'subtract 3'.

Example 2   Identify the mapping shown in this mapping list:

$0 \rightarrow 0$
$1 \rightarrow 2$
$2 \rightarrow 4$
$3 \rightarrow 6$

The mapping is: $x \rightarrow 2x$

Each time the number is being doubled. The rule is 'multiply by 2'.

Exercise 7.2

1 Identify the mappings on each of the following pairs of number lines:

a)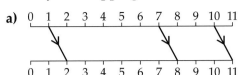

b)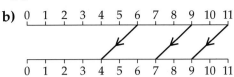

2 Identify the mappings on each of the following pairs of number lines:

a)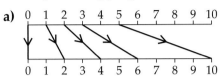

b)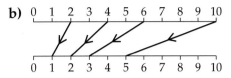

c)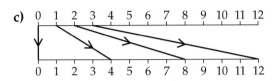

d)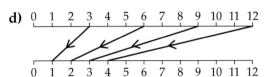

3 Identify the mapping for each of the following mapping lists:

a) $5 \rightarrow 15$
    $6 \rightarrow 16$
    $7 \rightarrow 17$

b) $1 \rightarrow 0$
    $2 \rightarrow 1$
    $3 \rightarrow 2$

c) $1 \rightarrow 16$
    $2 \rightarrow 17$
    $3 \rightarrow 18$

d) $11 \rightarrow 0$
    $22 \rightarrow 11$
    $33 \rightarrow 22$

4 Identify the mapping for each of the following sets of numbers:

a) $5 \rightarrow 20$
    $6 \rightarrow 24$
    $7 \rightarrow 28$

b) $0 \rightarrow 0$
    $5 \rightarrow 1$
    $10 \rightarrow 2$

c) $1 \rightarrow 12$
    $2 \rightarrow 24$
    $3 \rightarrow 36$

d) $10 \rightarrow 1$
    $100 \rightarrow 10$
    $1000 \rightarrow 100$

5 Write down two different mappings that could map:

a) $1 \rightarrow 5$

b) $3 \rightarrow 1$

c) $12 \rightarrow 2$

d) $8 \rightarrow 16$

6 a) Sketch mapping diagrams to show the following mappings.
    Draw the number lines from 0 to 20.

   i) $x \rightarrow x + 4$ **ii)** $x \rightarrow x + 1$ **iii)** $x \rightarrow x + 12$

   b) What do you notice about the arrows you have drawn on your mapping diagrams?
   Explain why this happens.

Special graphs

◈ Draw graphs of the form $y = a$ and $x = a$ when a is a number

You plot points on a **coordinate** grid using coordinates.

You write the distance **across** from the **origin** first. This is called the x-coordinate.

It gives the distance from the origin in the direction of the **x-axis**.

You write the distance **up** or **down** from the origin second. This is called the y-coordinate. It gives the distance from the origin in the direction of the **y-axis**.

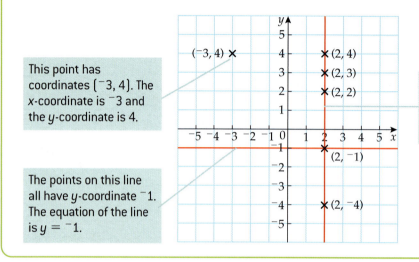

This point has coordinates ($^-$3, 4). The x-coordinate is $^-$3 and the y-coordinate is 4.

The points on the line all have x-coordinate 2. The equation of the line is $x = 2$.

The points on this line all have y-coordinate $^-$1. The equation of the line is $y = ^-1$.

Example 1   Draw the graph of $y = 2$.

The points on the line $y = 2$ all have a y-coordinate of 2.

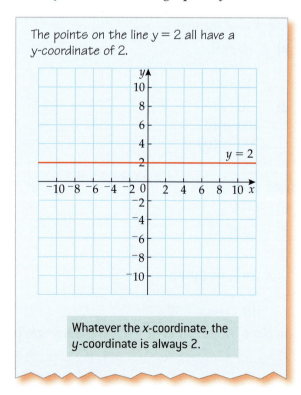

Whatever the x-coordinate, the y-coordinate is always 2.

Example 2   Find the equation of this line.

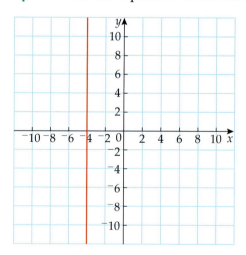

For every point on the line, the x-coordinate is -4.

The equation of the line is $x = ^-4$.

Whatever the y-coordinate, the x-coordinate is $^-4$.

Exercise 7.3

1 Copy this pair of axes onto squared paper.
Draw the following graphs on the pair
of axes. Label each graph with its equation.

a) $x = 3$ b) $y = 2$

c) $x = {}^-2$ d) $y = 0$

e) $x = 0$ f) $y = {}^-1.5$

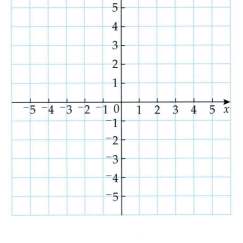

2 Look at the graphs you drew in **Q1**.

a) What do you notice about the graphs
with equations $x = \square$?
Copy and complete:
Graphs of the form $x = \square$ are parallel to
the …-axis

b) What do you notice about the graphs with
equations $y = \square$?
Copy and complete: Graphs of the form $y = \square$ are parallel to the …-axis

3 a) Draw a pair of axes labelled from ⁻5 to 5 on squared paper.

> You could copy the axes from **Q1**.

b) Plot the points (2, 3), (2, ⁻1), (2, 4), (2 ⁻5)

c) Join the points with a straight line.

d) What is the equation of the line?

e) Explain how you could work this out without drawing
the graph.

> Look at the coordinates.

4 Write down the equations of the lines which go through
the following pairs of coordinates.

> You may wish to plot the points and draw a line through them.

a) (1, 7) and (1, ⁻4)   b) (2, 3) and (4, 3)

c) (⁻1, 1) and (⁻1, 0)   d) (⁻5, ⁻1.5) and (3, ⁻1.5)

5 a) Which of the following lines are symmetrical
about the x-axis? For example, $x = {}^-4$ is
symmetrical about the x-axis:

$y = 7$ $x = 4$ $y = {}^-3$ $x = {}^-1$
$x = 5$ $y = 3$ $y = {}^-2$ $x = {}^-5$

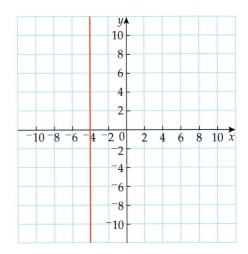

b) Write down the equations of the lines that
are symmetrical about the y-axis.

c) In general, which lines are symmetrical
about the x-axis?

d) In general, which lines are symmetrical
about the y-axis?

6 a) What is the equation of the x-axis?

b) What is the equation of the y-axis?

> What can you say about the y-coordinate
when you are on the x-axis?

Graphs

◈ Plot a straight-line graph from its equation

◈ Read and interpret straight-line graphs

A **graph** is a set of **points** connected by a rule. You can write the rule as an **equation**, for example $y = 3x$, $y = x + 1$.

The **equation** tells us about the relationship between the x- and y-values.

You can draw a table to show the relationship between the x- and y-values.

For example, for the graph $y = 2x$:

x	⁻1	0	1	2
y	⁻2	0	2	4
Coordinates of point	(⁻1, ⁻2)	(0, 0)	(1, 2)	(2, 4)

When $x = 2$,
$y = 2 \times 2 = 4$

1) Choose the value of an x-coordinate.

2) Find the y-coordinate for this x-coordinate, using the equation.

3) Plot the pair of coordinates (x, y)

4) Plot some more points and then join them with a straight line.

Example   Plot the graph of $y = x - 2$

First draw up a table to show the x- and y-values for $y = x - 2$.
When $x = 0$, $y = 0 - 2 = ⁻2$ When $x = 1$, $y = 1 - 2 = ⁻1$
When $x = 2$, $y = 2 - 2 = 0$ When $x = 3$, $y = 3 - 2 = 1$

x		0	1	2	3
y		⁻2	⁻1	0	1
Coordinates of point		(0, ⁻2)	(1, ⁻1)	(2, 0)	(3, 1)

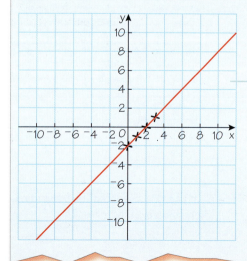

Next draw a pair of axes, plot the points and join them with a straight line.

Exercise 7.4

1 On squared paper, draw a pair of axes labelled from $^-10$ to 10.
Copy and complete the tables, then draw the graph for each one on the same pair of axes.

a) $y = 3x$

x	0	1	2	3
y	$3 \times 0 = 0$			

b) $y = 4x$

x	0	1	2	3
y				

2 a) Copy and complete the table below for: $y = x + 2$

x	0	1	2	3	4
y	$0 + 2 =$	$1 + 2 =$			

b) On squared paper, draw a pair of axes labelled from $^-10$ to $^+10$.
c) Plot the points from your table on your axes.
d) Join the points with a straight line. Continue it as far as possible.

3 Repeat **Q2** for the graph of $y = x - 1$.

4 On squared paper, draw a pair of axes labelled from $^-10$ to $^+10$.
a) Draw up a table of values for each of these graphs:
    i) $y = x$ **ii)** $y = x + 1$ **iii)** $y = x + 3$
b) Plot the three graphs from part **a)** on your axes. Label each graph with its equation.
c) Describe what you notice about the graphs.

> Will the graphs ever cross each other?

5 Look at your graphs from **Q4**.
a) Copy and complete:
The graph of $y = x$ crosses the y-axis at $y = \square$
The graph of $y = x + 1$ crosses the y-axis at $y = \square$
The graph of $y = x + 3$ crosses the y-axis at $y = \square$
b) Where do you think the graph of $y = x + 5$ will cross the y-axis?
c) Draw the graph of $y = x + 5$ to check your answer to part **b)**.

> Look at the pattern in your answers to part **a)**.

6 a) Copy and complete the table below for: $y = 2x + 1$

x	0	1	2	3	4
y		$2 \times 1 + 1$ $= 2 + 1$ $= 3$			

b) Draw a pair of axes from $^-15$ to $^+15$.
c) Plot the points you have calculated in part **a)** on your axes.
d) Join the points with a straight line. Continue it as far as possible.

7 Repeat **Q6** for the graph of $y = 3x + 2$.

The *y*-intercept

⊕ Find where a graph crosses the *y*-axis (the *y*-intercept)

The point where a graph crosses the *y*-axis is called the **y-intercept** .
For example, for the graph $y = 2x + 3$:

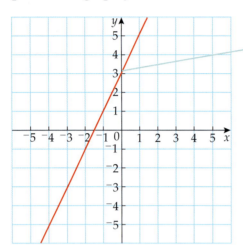

This is the *y*-intercept.

The *x*-coordinate of the *y*-intercept is always 0, because the equation of the *y*-axis is $x = 0$.

All **equations** of **straight-line graphs** can be written in the form: $y = mx + c$ where *m* and *c* are numbers. The *y*-intercept of the graph $y = mx + c$ is $(0, c)$.

Example   What are the coordinates of the *y*-intercept of the graph $y = 3x - 4$?

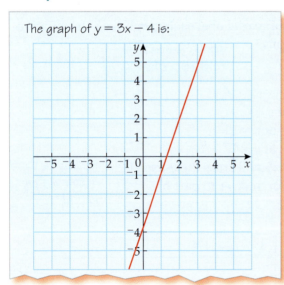

The graph of $y = 3x - 4$ is:

The *y*-intercept is the point: $(0, ^-4)$.

You could work out the *y*-intercept without drawing the graph. Look at the equation $y = 3x - 4$. The '−4' tells you that the coordinates of the *y*-intercept are $(0, ^-4)$.

Exercise 7.5

You may use a graph sketching package to check your answers to the following questions.

1. Write down the coordinates of the *y*-intercept of each of the following graphs:
   a) $y = x + 5$
   b) $y = 3x + 5$
   c) $y = x + 1$
   d) $y = x + 7$
   e) $y = 7x + 1$
   f) $y = 8x + 6$
   g) $y = 10x + 0.5$
   h) $y = 4x + 2$
   i) $y = x + 11$

2 Write down the coordinates of the y-intercept of each of the following graphs:

a) $y = x - 1$ b) $y = x - 2$ c) $y = 2x - 1$

d) $y = 7x - 5$ e) $y = 2x - 2$ f) $y = 8x - 4$

g) $y = 9x - 3$ h) $y = 10x - 1.5$ i) $y = 7x - 3.5$

3 Write down the equation of two different graphs whose y-intercept is:

> Your equation should be of the form $y = mx + c$, where m is a number and c is the y-intercept.

a) $(0, 5)$ b) $(0, 11)$

c) $(0, {}^-5)$ d) $(0, {}^-2.5)$

e) $(0, {}^-1)$ f) $(0, 9.9)$

g) $(0, \frac{1}{4})$ h) the origin

4 Match the equations of the lines with their graphs:

i) 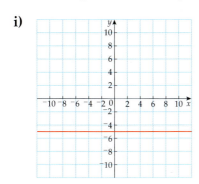   ii) 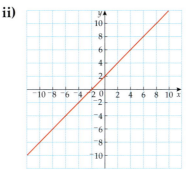   iii)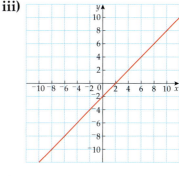

a) $y = x + 2$ b) $y = {}^-5$ c) $y = x - 2$

5 A mobile telephone company draws a graph using the equation:

$y = 0.5x + 10$ where y = monthly bill in £ and x = number of minutes of calls.

a) What is the y-intercept of this graph?
b) What does the y-intercept represent?
c) What is the monthly line rental?
d) How much is the monthly bill for 1 minute of calls?
e) How much is the monthly bill for 20 minutes of calls?
f) What is the cost per minute for calls?

6 a) Copy and complete the table below:

Graph	y-intercept
$y = 2x + 5$	$(0, \quad)$
$y = 5 + 2x$	
$y = 3x - 2$	
$y = -2x + 3$	
$y = 10x - 1$	
$y = 1 - 10x$	

b) Where do you think the graph of $y = 13 + 2x$ will intercept the y-axis?

Gradients

⊕ Understand what is meant by the gradient of a line

⊕ Recognise parallel lines from their equations

The **gradient** of a hill is a measure of how steep it is.

In maths we talk about the **gradient** of a line.

The equation of a straight line can be written in the form:

$y = mx + c$ where m and c are numbers.

m is the gradient and c is the y-intercept.

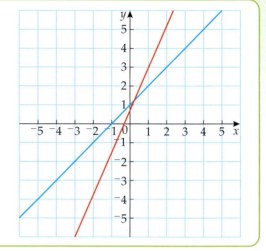

> The red line is steeper than the blue line. We say that the **gradient** of the red line is **steeper** than the **gradient** of the blue line.

Example   a) Draw the graphs $y = 3x$ and $y = 3x + 2$ on the same pair of axes.
  b) Describe the graphs.
  c) Where does each graph cross the y-axis?

a) $y = 3x$

x	0	1	2	3
y	0	3	6	9

$y = 3x + 2$

x	0	1	2	3
y	2	5	8	11

Draw up tables of values.

b) The graphs of $y = 3x$ and $y = 3x + 2$ are   **parallel** .
c) $y = 3x$ crosses the y-axis at the origin
    $y = 3x + 2$ crosses the y-axis at (0, 2)

In the equation $y = 3x$, $c = 0$

In the equation $y = 3x + 2$, $c = 2$

Exercise 7.6

You may use a graph sketching package for this exercise.

1 Plot the following graphs on the same pair of axes, labelled from $^-10$ to $^+10$ on both axes.

    a) $y = x$
    b) $y = 2x$
    c) $y = 4x$

    d) Describe what happens to the gradient of the graphs as the number in front of the x increases.

2 **a)** Draw the graphs of $y = 3x$ and $y = 5x$.

> You could use your axes from **Q1**.

    b) Which has the steepest gradient?

    c) Write down the equation of a line that has a steeper gradient than $y = 5x$.

> You may wish to check your answer by drawing.

3 **a)** Plot these graphs on the same pair of axes, labelled from $^-10$ to $^+10$ on both axes.

      i) $y = 2x + 2$
      ii) $y = 2x$
      iii) $y = 2x - 3$

    b) Does the value of the y-intercept affect the gradient of the graph?

    c) Copy and complete:

      Graphs are _____ when the number in front of the x in their equations is the same.

    d) Copy and complete:

      $y = 2x + 2$ is parallel to $y = \square x$, $y = \square x + 5$, $y = \square x - 13$, $y = \square x + \square$

4 Find pairs of graphs which are parallel.

    $y = 2x + 3$ $y = 6x - 1$ $y = 2x - 7$ $y = 5x + 2$ $y = 6x + 7$

    $y = 4x - 4$ $y = 4x$ $y = x$ $y = 5x - 5$ $y = x + 1$

5 Write down the equation of a line parallel to:

> Look at your answer to **Q3** part **c)**.

    a) $y = 2x$
    b) $y = 7x$
    c) $y = 3x$

    d) $y = 5x + 1$
    e) $y = 4x - 5$
    f) $y = 11x - 11$

6 **a)** Draw the graphs of $y = 3x$, $y = 2x$, $y = x$ and $y = 0.5x$

    b) What happens as the number in front of the x gets smaller?

    c) Write down the equation of a graph with a gradient that is less steep than $y = 0.5x$.

7 Match the equations of the lines with the graphs:

> Remember the point at which the graph crosses the y-axis is given by the value of c when $y = mx + c$.

    a) $y = 2x$
    b) $y = 3x$
    c) $y = 2x + 3$

    i)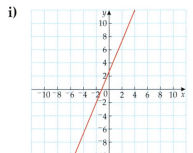
    ii)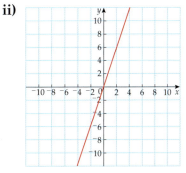
    iii)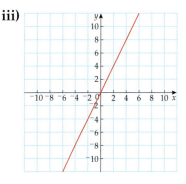

Converting between metric units

◈ Convert between metric units of length, mass and capacity

Metric units of **length** are: kilometre (km), metre (m), centimetre (cm) and millimetre (mm).

10 mm = 1 cm
100 cm = 1 m
1000 m = 1 km

Metric units of **mass** are: kilogram (kg) and gram (g).

1000 g = 1 kg

Capacity is a measure of the amount a container will hold. Metric units of capacity are: litre (ℓ), centilitre (cl) and millilitre (ml).

10 ml = 1 cl
100 cl = 1 ℓ
1000 ml = 1 ℓ

Example 1 Convert:

a) 2 km to m
b) 300 g to kg
c) 420 ml to cl

a) $2 \times 1000 = 2000$ m — To convert km to m multiply by 1000.

b) $300 \div 1000 = 0.3$ kg — To convert g to kg divide by 1000.

c) $420 \div 10 = 42$ cl — To convert ml to cl divide by 10.

Example 2 A lorry carries 200 planks of wood. Each plank is 8.5 m long and weighs 80 kg.
It takes a forklift truck 2 minutes to unload 10 planks.

a) What is the total length of the planks on a lorry?
b) What is the total weight of the planks?
c) How long does it take to unload the lorry?

a) Total length = $8.5 \times 200 = 1700$ m — Multiply the length of one plank (8.5 m) by the total number of planks (200).

b) Total weight = $200 \times 80 = 16\,000$ kg — Multiply the weight of one plank (80 kg) by the total number of planks (200).

c) $200 \div 10 = 20$ loads
$20 \times 2 = 40$ minutes — It takes the truck 2 minutes to unload 10 planks. There are 20 lots of 10 planks in the lorry so it takes $20 \times 2 = 40$ minutes to unload them all.

Exercise 8.1

1 Convert the following measurements into the units stated:

a) 3 cm into mm   b) 1.1 m into mm

c) 0.4 cm into mm   d) 30 mm into cm

e) 5.1 m into cm   f) 3000 m into km

g) 2000 mm into m   h) 250 cm into m

i) 0.3 km into m   j) 4.5 km into cm

k) 1 000 000 mm into km   l) 2.5 km into mm

To convert from	to	you need to
km	m	× 1000
m	km	÷ 1000
m	cm	× 100
cm	m	÷ 100
km	m	× 1000
m	km	÷ 1000

2 Put the following in order, smallest to largest:

a) 3.6 kg, 1500 g, $\frac{1}{2}$ kg, 800 g, $\frac{7}{8}$ kg

b) 3.3 kg, 3200 g, $3\frac{1}{2}$ kg, 3450 g, 3400 g

Write all the masses in a common unit to compare them.

To convert from	to	you need to
kg	g	× 1000

3 Convert the following measurements into the units stated:

a) 3 ℓ into cl   b) 2.6 ℓ into cl

c) 350 ml into cl   d) 2.4 ℓ into ml

e) 35 cl into ml   f) 1.75 ℓ into ml

g) 500 ml into ℓ   h) 150 cl into ℓ

i) 65.5 cl into ℓ

To convert from	to	you need to
ℓ	ml	× 1000
ml	ℓ	÷ 1000
cl	ml	× 10
ml	cl	÷ 10

4 Find the size of the missing lengths.

a)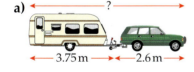

b)

c)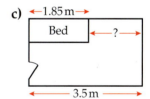

5 A coach company expects its coaches to travel about 240 miles each day.
Each coach travels for about 350 days a year.

Multiply the number of miles the coach travels each day by the number of days the coach travels in a year.

a) Approximately how many miles will a coach travel in a year?

b) The company replaces its coaches every 8 years. Approximately how many miles will a coach travel before it is replaced?

6 The instructions for roasting a chicken are:

30 minutes per 450 g plus an extra 20 minutes.

Aaron has a 900 g chicken.

a) How long should he cook the chicken?

b) He wants it to be ready at 13 00. What time should he put it in the oven?

It takes 15 minutes for the oven to warm up. What is the latest time he should put the oven on?

Ordering decimals

Key words
decimal number
digit
units
tenths
hundredths

◈ Order decimals by comparing digits
◈ Order decimals by positioning them on a number line

When we read a **decimal number**, we have to imagine the headings above the **digits**:

T	U	.	t	h
3	5	.	8	2

The decimal point separates the whole numbers from the parts of whole numbers.

In the decimal number 35.82

the value of the digit '3' is 3 tens

the digit '5' is 5 **units**

the digit '8' is 8 **tenths**

the digit '2' is 2 **hundredths**

To put decimal numbers in order, look at the whole number parts first. If they are the same, compare the tenths digits. If the tenths digits are the same, compare the hundredths digits, and so on. For example, 4.27 is greater than 4.21 because $\frac{7}{100}$ is greater than $\frac{1}{100}$.

Example 1 Put these numbers in order from smallest to greatest:

3.25   3.245   3.193   3.3

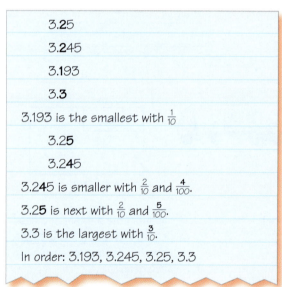

3.25

3.245

3.193

3.3

3.193 is the smallest with $\frac{1}{10}$

3.25

3.245

3.245 is smaller with $\frac{2}{10}$ and $\frac{4}{100}$.

3.25 is next with $\frac{2}{10}$ and $\frac{5}{100}$.

3.3 is the largest with $\frac{3}{10}$.

In order: 3.193, 3.245, 3.25, 3.3

> All units are the same so we look at the tenths column.

> These have the same units and the same tenths so we look at the hundredths column.

Example 2 Place these decimals on a number line and then write them in order from smallest to greatest:

2.163   2.251   2.13   2.28   2.214

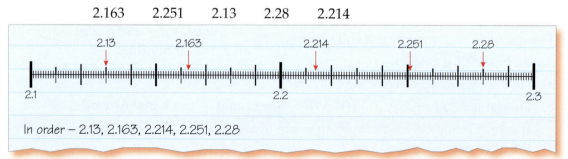

In order – 2.13, 2.163, 2.214, 2.251, 2.28

Exercise 8.2

1 132.675

What is the value of:

a) 6 in this number   **b)** 3 in this number   **c)** 7 in this number

d) 2 in this number   **e)** 5 in this number   **f)** 1 in this number?

2 Place these in order from smallest to greatest.

a) 0.3 0.5 0.1 0.4 0.2   **b)** 2.36 2.32 2.37 2.3 2.34

c) 0.247 0.249 0.243 0.242 0.24   **d)** 1.8 1.862 1.86 1.869

3 Place these numbers on a number line, then write them in order from smallest to greatest.

a)

```
 |-+-+-+-+-+-+-+-+-+-|
 3                   4
```
3.4 3.7 3.2 3.5 3.8

b)

```
 |-+-+-+-+-+-+-+-+-+-|
 6.1               6.2
```
6.17 6.15 6.13 6.14 6.18

c)

```
 |⊢⊢⊢⊢⊢⊢⊢⊢⊢⊢⊢⊢⊢⊢⊢⊢⊢⊢⊢⊢|
 5.2                  5.3
```
5.22 5.26 5.213
5.272 5.255

4 Write down the larger number in each pair:

a) 1.6 2.1   **b)** 3.26 3.24   **c)** 6.725 7.722

d) 4.3 4.29   **e)** 8.299 8.3   **f)** 16.09 16.1

5 These are the times for the boys' 200 m final. Put them in order, first to fifth.

Tom White   47.47 secs

John Brown   48.99 secs

Billy Grey   47.60 secs

Jim Green   48.18 secs

Fred Black   47.46 secs

6 True or False?

a) $0.24 < 0.243$ **b)** $6.254 > 6.26$ **c)** $1.359 > 1.299$

d) $5.86 < 5.859$ **e)** $9.23 < 9.3$ **f)** $7.452 > 7.449$

> Remember $<$ means 'less than' and $>$ means 'more than'.

7 Which number is halfway between:

a) 5 and 7   **b)** 12 and 13   **c)** 2.5 and 2.3   **d)** 6.122 and 6.126?

8 A game for two players.

Draw a box like this:

On another piece of paper, write down four numbers between 2 and 3 with up to 3 d.p. Do not let your partner see them. Take turns to write one number on the grid, starting with the middle cell.

The numbers must be ordered correctly.

The winner is the person who is able to write more numbers on the grid.

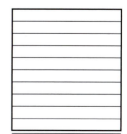

Rounding

- Round whole numbers to the nearest 10, 100, 1000, …
- Round decimal numbers to the nearest whole number, tenth or thousandth

Key words
round
round up
round down
nearest tenth
nearest hundredth

4637 lies between the thousands 4000 and 5000 – its nearest thousand is 5000

> The halfway point is 4500.

4637 lies between the hundreds 4600 and 4700 – its nearest hundred is 4600

> The halfway point is 4650.

4637 lies between the tens 4630 and 4640 – its nearest ten is 4640

> The halfway point is 4635.

Finding the nearest thousand, hundred, ten and so on is called **rounding** . We **round up** 4637 to 5000 (to the nearest thousand). We **round down** 4338 to 4000 (to the nearest thousand).

If a number lies exactly halfway between two numbers, then we round up.

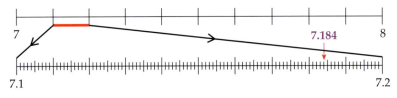

7.184 lies between the whole numbers 7 and 8 – its nearest whole number is 7

> The halfway point is 7.5.

7.184 lies between the tenths 7.1 and 7.2 – its nearest tenth is 7.2

> The halfway point is 7.15.

7.184 lies between the hundredths 7.18 and 7.19 – its nearest hundredth is 7.18

> The halfway point is 7.185.

Rounding to the **nearest tenth** is also called rounding to one decimal place.

Rounding to the **nearest hundredth** is also called rounding to two decimal places.

Example 1 Round this number to the nearest **a)** 10 **b)** 100 **c)** 1000

6537

> a) To the nearest 10   6537 → 6540
> b) To the nearest 100   6537 → 6500
> c) To the nearest 1000 6537 → 7000

> 65**3**7 is between 6530 and 6540. It is more than halfway (6535).

> 6**5**37 is between 6500 and 6600. It is less than halfway (6550).

> **6**537 is between 6000 and 7000. 6500 is halfway so we round up.

Example 2 Round 5.375 to:

 a) 1 decimal place   **b)** 2 decimal places

a) To 1 decimal place (the nearest tenth)
5.375 → 5.4
b) To 2 decimal places (the nearest hundredth)
5.375 → 5.38

5.**3**75 is between 5.3 and 5.4. Halfway is 5.35 so round up.

5.3**7**5 is between 5.37 and 5.38. Halfway is 5.375 and our number is exactly halfway. In this case we round up to 5.38.

Exercise 8.3

1 Round each price to the nearest pound.
 a) £3.26   **b)** £7.39   **c)** £2.52   **d)** £17.90   **e)** £36.50

2 Round these ages to the nearest 10 years.
 a) 54   **b)** 62   **c)** 39   **d)** 102   **e)** 5

3 These are the populations of different towns in England. Round them to the nearest 1000.
 a) 53 850   **b)** 163 490   **c)** 274 499   **d)** 93 570   **e)** 384 700

4 Which number is halfway between:
 a) 5 and 6   **b)** 3.1 and 3.2   **c)** 6.5 and 6.6
 d) 4.27 and 4.28   **e)** 9.32 and 9.33   **f)** 8.844 and 8.845?

5 Round these heights to 1 decimal place (1 d.p.).
 a) 134.49 cm   **b)** 148.61 cm   **c)** 154.53 cm   **d)** 149.35 cm   **e)** 162.27 cm

6 Round these javelin throws to 2 d.p. (2 decimal places).
 a) 53.629 m   **b)** 57.246 m   **c)** 56.632 m
 d) 54.365 m   **e)** 57.251 m

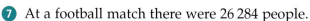

7 At a football match there were 26 284 people.
 a) How many is this to the nearest ten thousand?
 b) How many is this to the nearest thousand?
 c) How many is this to the nearest hundred?
 d) How many is this to the nearest ten?

8 Round each number to the nearest whole number and then make an estimate of each calculation.
 a) 6.2×2.3 **b)** $4.3 \div 1.9$ **c)** $9.8 \div 2.4$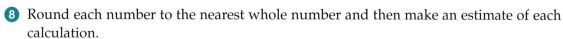

9.8 rounds to 10; 2.4 rounds to 2; $10 \times 2 = 20$.

 d) 3.17×4.89 **e)** $3.85 + 4.14 + 8.71$

9 List all the numbers with 1 decimal place, that round to 7 when rounding to the nearest whole number. Use this number line to help you.

10 Use the digits 5, 2, 3 and 8.
 Make numbers with 3 decimal places, for example 5.238
 You may use only one of each digit in any number.
 Suggest one number for each of these answers:
 a) 3.5 rounded to the nearest 1 decimal place
 b) 2 rounded to the nearest whole number
 c) 3.29 rounded to 2 decimal places.

Adding and subtracting

Key words
standard method
estimate
digit
decimal point

◈ Add whole numbers and decimal numbers using a standard written method

◈ Subtract whole numbers and decimal numbers using a standard written method

You can add and subtract using **standard methods** .

Start by writing an **estimate** of the answer by rounding.

Write the numbers in columns underneath each other, making sure that each **digit** is in its correct column and that the **decimal points** are lined up.

To add:
Start adding from the right, carrying into the next column on the left, if necessary.

To subtract:
Start subtracting from the right, 'requesting' a digit from the next column on the left, if necessary.

Finally, compare the answer with the estimate as a check.

$$
\begin{array}{r}
1\,7\,.\,3\,6 \\
8\,.\,9 \\
+\ 4\,2\,.\,1\,5 \\
\hline
6\,8\,.\,4\,1 \\
\scriptstyle 1\ \ 1\ \ \ \ 1
\end{array}
\qquad
\begin{array}{r}
\scriptstyle 6 \ \ \ 12\ 1 \\
2\,\not{7}\,.\,\not{3}\,5 \\
-\ 5\,.\,6\,9 \\
\hline
2\,1\,.\,6\,6
\end{array}
$$

Example 1   Work out $17.23 + 3.4 + 2.49$

Round to the nearest whole number.

Estimate: $17 + 3 + 2 = 22$

$$
\begin{array}{r}
1\,7\,.\,2\,3 \\
+\ \ \ \ 3\,.\,4 \\
2\,.\,4\,9 \\
\hline
2\,3\,.\,1\,2 \\
\scriptstyle 1\ \ 1\ \ \ \ 1
\end{array}
$$

Set them out with the decimal points lined up.

Check your estimate: 23.12 is close to 22 ✓

Example 2   Find the difference between 43.8 and 13.37

Estimate: $44 - 13 = 31$

Fill the space with 0.

$$
\begin{array}{r}
\scriptstyle 7 \ \ 1 \\
4\,3\,.\,\not{8}\,0 \\
-\ 1\,3\,.\,3\,7 \\
\hline
3\,0\,.\,4\,3
\end{array}
$$

To subtract 7 from 0, we need to request a number from the next column.

Check estimate: 30.43 is close to 31 ✓

Exercise 8.4

1 Complete these additions. Remember to estimate first and then check your answer.

   a) $4.06 + 3.7 + 6.1$ **b)** $2.95 + 4.3 + 0.62$ **c)** $9.3 + 2.85 + 7.07$

   d) $15.6 + 3.795 + 4.28$ **e)** $12.32 + 6.97 + 5$

2 Complete these subtractions:

   a) $4.75 - 2.36$ **b)** $9.2 - 4.11$ **c)** $11.32 - 6.52$

   d) $12.43 - 5.7$ **e)** $8.3 - 1.87$

3 Copy and complete this addition pyramid.
   Each brick is found by adding the two directly below it.

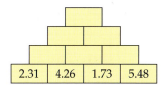

| 2.31 | 4.26 | 1.73 | 5.48 |

4 Copy and complete this subtraction pyramid.
   Each brick is found by finding the difference between
   the two directly below it. (Take the smaller number away
   from the larger number.)

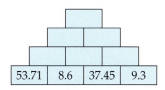

| 53.71 | 8.6 | 37.45 | 9.3 |

5 A pack of toffee sweets costs £1.37, a pack of chews
   costs £1.05 and a box of chocolates costs £3.75.
   How much is this altogether?

6 Peter had £46.50 birthday money. He spent £39.75 on a computer game.
   How much does he have left?

7 Copy the boxes below.

   □□.□□□ + □□.□□□ + □□.□□□

   Use 0–9 digit cards. Shuffle the cards and place one in each box to make numbers with
   two decimal places. Estimate the answer to the addition sum you make.
   Now add the three numbers together.
   Play three rounds then add your answers together. What is their total?
   Round this to 1 decimal place.

Investigation

8 Using the boxes and cards in **Q7**, make three numbers that will give you

   a) the greatest possible answer

   b) the smallest possible answer.

Multiplying

⊕ Multiply a 3-digit number by a 2-digit number using a standard method
⊕ Estimate the result of a multiplication by rounding

You can use different methods to multiply numbers. Here are two ways of calculating 258×24.

A grid method:

	200	50	8
20	4000	1000	160
4	800	200	32

```
  5 1 6 0
+ 1 0 3 2
  ───────
  6 1 9 2
```

A **standard method** :

```
    2 5 8
  ×   2 4
  ───────
      3 2
    2 0 0
    8 0 0
    1 6 0
  1 0 0 0
  4 0 0 0
  ───────
  6 1 9 2
  ₁
```

or

```
      2 5 8
    ×   2 4
    ───────
  1 ₀0 ₃3 2
    5 ₁1 6 0
    ───────
    6 1 9 2
```

When multiplying numbers you should always:
• **Estimate** the answer by rounding then multiplying the rounded numbers.
• Work out the multiplication.
• Check that your answer is sensible by comparing it with your estimate.

Example Find the total cost of a visit to the theatre for 123 people if tickets cost £32 each.

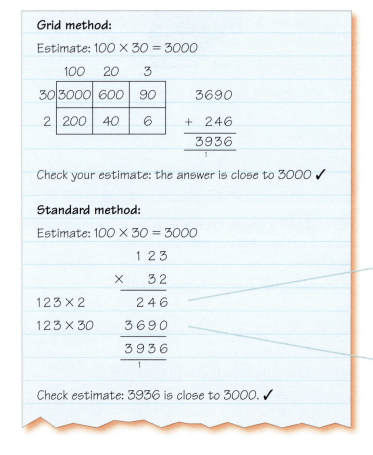

Round three digits to the nearest 100 and two digits to the nearest 10.

Grid method:

Estimate: $100 \times 30 = 3000$

	100	20	3
30	3000	600	90
2	200	40	6

```
    3690
+   246
  ──────
    3936
      1
```

Check your estimate: the answer is close to 3000 ✓

Standard method:

Estimate: $100 \times 30 = 3000$

```
          1 2 3
      ×     3 2
      ─────────
123 × 2     2 4 6
123 × 30  3 6 9 0
          ─────────
          3 9 3 6
              1
```

Check estimate: 3936 is close to 3000. ✓

```
  123
×   2
────
 246
```

$123 \times 30 = 123 \times 10 \times 3$
$123 \times 10 = 1230$
```
       × 3
      ────
      3690
```

Exercise 8.5

1 Copy and complete these number sentences:

a) $426 \times 20 = 426 \times 10 \times 2 = 4260 \times \square$

b) $539 \times 50 = \square \times 10 \times 5 = 5390 \times 5$

c) $653 \times 70 = 653 \times \square \times 7 = \square \times 7$

2 Now complete the multiplications from **Q1**.

3 Complete these multiplications:

Remember to estimate your answer first.

a) 321×23 b) 432×31

c) 673×25 d) 437×53

4 Copy and complete these multiplication pyramids. Each brick is found by multiplying the two directly below it.

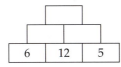
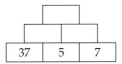

5 There are 225 beads in a packet. Mrs Khan bought 23 packets. How many beads did she have altogether?

6 It costs 32p for a notebook.
Tipton High School buys 875 of them.
How much is this?

You can give your answer in £ and pence or just pence, but £ and pence is more usual.

7 To lay a path Wasim needs 347 bricks.
Each brick is 14 cm long.
How long is the path in centimetres?
How long is this in metres?

100 cm = 1 m.

8 Use 0–9 digit cards. Shuffle the cards and make a 3-digit number and a 2-digit number.
Now multiply them together, and write down the answer.
Repeat this several times.

Investigation

9 Use the four consecutive digit cards 5, 6, 7 and 8.

Create two 2-digit numbers, using each digit only once.

Multiply them together. Investigate which two numbers give the largest possible answer and which give the smallest possible answer.

Repeat the activity for a different set of five consecutive digits.

Can you see a pattern?

Challenge – make your five digits into a 3-digit number and a 2-digit number. Now can you find the largest possible answer and the smallest possible answer?

More multiplying

⊕ Multiply a 1- or 2-place decimal number by a 1-digit whole number using a standard method

⊕ Estimate the result of a multiplication by rounding

To multiply a decimal number using a **standard written method** :

Step 1: **Estimate** the answer by rounding, then multiply the rounded numbers.

Step 2: Change the decimal number into a whole number by multiplying by 10 or 100:
if it has one decimal place, multiply it by 10
if it has two decimal places, multiply it by 100.

Step 3: Work out the multiplication.

Step 4: Divide the answer by 10 or 100 (whatever you multiplied by in Step 2).

Step 5: Check that your answer is sensible by comparing it with your estimate.

For example to calculate: 4.76×6
Multiply by 100 : 476×6

longer method

```
    4 7 6
×       6
    3 6
  4 2 0
2 4 0 0
2 8 5 6
```

shorter method

```
    4 7 6
×       6
  2 8 5 6
    4 3
```

Divide by 100 : $2856 \div 100 = \underline{28.56}$

Example 1 Work out: 8.3×7

Round 8.3 to the nearest whole number (8).

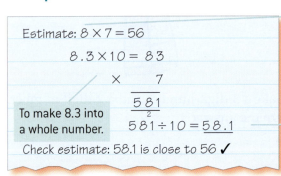

Estimate: $8 \times 7 = 56$

$8.3 \times 10 = 83$

```
    ×     7
    5 8 1
      2
```

To make 8.3 into a whole number.

$581 \div 10 = \underline{58.1}$

$\div 10$ to 'undo' the $\times 10$ we did at the beginning.

Check estimate: 58.1 is close to 56 ✔

Example 2 Work out the cost of eight games at £9.36 each.

Estimate: $8 \times £9 = £72$

Rounding £9.36 to the nearest whole number (9).

$9.36 \times 100 = 936$

```
      ×   8
    7 4 8 8
      2 4
```

$\times 100$ to make 9.36 into a whole number.

$\div 100$ to 'undo' the $\times 100$

$7488 \div 100 = £74.88$

Check estimate: £74.88 is close to £72 ✔

Exercise 8.6

1 Work out:

a) 4.2×4 b) 6.32×3 c) 5.19×5 d) 8.9×9 e) 3.85×6

2

Jayne needs four pieces of wood 1.75 metres long for her garden fence.

a) How much wood is this altogether?

b) The wood costs £3.28 per metre.

How much does Jayne's wood cost?

3 Calculate the total cost of the following:

a) Socks: 3 pairs at £1.23 per pair

b) Hair slides: 5 at £0.65 each

c) Scarf: 6 at £2.34 each

d) Gloves: 3 pairs at £3.58 per pair

e) Magazines: 7 at £1.49 each

4 Copy and complete this multiplication chain.

3.21   ×3   ×5   ×2

5 Use a dice to throw three numbers. Write them in the form $\Box . \Box \Box$.

Now place them in the multiplication chain in **Q4** and work them out.

Write your final answer correct to 1 d.p.

Do this three times.

6 Use 0–9 digit cards with their numbers showing.

Pick three cards to make a number with 2 decimal places and one number to multiply by.

Your target is to make an answer as close as you can to 14.

Write down each answer you get and round it to the nearest whole number.

You score 20 points for getting 14 exactly and 5 points for each answer that rounds to 14.

Complete five rounds and find your total score.

Multiples

⊕ Recognise and use multiples

⊕ Find the lowest common multiple of two numbers

Key words
multiple
common multiple
lowest common multiple

The **multiples** of a number are the numbers in its multiplication table.

For example, the multiples of 6 are:

6, 12, 18, 24, 30, 36, 42, …

The first ten multiples of the numbers from 1 to 10 appear in the rows and columns of a multiplication square.

The multiples of 3 are: 3, 6, 9, 12, 15, 18, 21, …

The multiples of 4 are: 4, 8, 12, 16, 20, 24, 28, …

	1	2	3	4	5	6	7	8	9	10
1	1	2	3	4	5	6	7	8	9	10
2	2	4	6	8	10	12	14	16	18	20
3	3	6	9	12	15	18	21	24	27	30
4	4	8	12	16	20	24	28	32	36	40
5	5	10	15	20	25	30	35	40	45	50
6	6	12	18	24	30	36	42	48	54	60
7	7	14	21	28	35	42	49	56	63	70
8	8	16	24	32	40	48	56	64	72	80
9	9	18	27	36	45	54	63	72	81	90
10	10	20	30	40	50	60	70	80	90	100

The numbers that appear in both lists are called the **common multiples** . The common multiples of 3 and 4 are: 12, 24, 36, 48, 60, …

The smallest of the common multiples is called the **lowest common multiple (LCM)** . The LCM of 3 and 4 is 12.

Example 1   a)   Find all the common multiples of 3 and 5, that are less than 50.

b)   Find the lowest common multiple (LCM) of 3 and 5.

> The multiples of 3 are: 3, 6, 9, 12, **15**, 18, 21, 24, 27, **30**, 33,
> 36, 39, 42, **45**, 48
> The multiples of 5 are: 5, 10, **15**, 20, 25, **30**, 35, 40, **45**
> a)   The common multiples are: 15, 30, 45
> b)   The LCM is 15

They appear in both lists of multiples.

Example 2   Two children were playing with toy drums. Tim banged his drum every 7 seconds and Sarah banged her drum every 3 seconds.

They started together, but when was the next time they both banged their drums together?

> John hit: 7, 14, **21**, 28, 35 seconds
> Sarah hit: 3, 6, 9, 12, 15, 18, **21**, 24 seconds
> They both banged their drum after 21 seconds.

Exercise 8.7

1 Find the first six multiples of:

    a) 4   **b)** 7   **c)** 8   **d)** 11   **e)** 25

2 Find all the common multiples, less than 25, of:

    a) 2 and 3   **b)** 3 and 4   **c)** 2 and 4

    d) 2 and 5   **e)** 3 and 5   **f)** 6 and 8

3 Use your answers to **Q2** to find the LCM of:

    a) 2 and 5   **b)** 3 and 4   **c)** 4 and 5

4 Find the common multiples less than 30 of the following pairs of numbers. Use your answers to find the LCM of each pair.

    a) 6 and 9   **b)** 2 and 7   **c)** 4 and 6

5 Copy and complete this LCM grid.

LCM	2	5	6
3	6		
4			
8			

The LCM of 2 and 3 is 6.

6 Spot barked every 3 seconds. Ben barked every 4 seconds.

Find how many seconds it takes for them both to bark at the same time.

List the times of each dog's barks.

7 Shep joins the other two dogs in **Q6**.
He barks every 5 seconds.

How many seconds will it be before all three dogs bark at the same time?

8 True or False?

    a) Multiples of an odd number are always odd.

    b) Multiples of 3 are also multiples of 6.

    c) Multiples of 10 are also multiples of 5.

    d) Common multiples of 3 and 4 are even numbers.

    e) Common multiples of 3 and 5 are all odd numbers.

9 Find the LCM of the denominators of these fractions.

    a) $\frac{2}{3} + \frac{3}{4}$ **b)** $\frac{5}{7} - \frac{1}{2}$

    c) $\frac{5}{6} - \frac{2}{5}$ **d)** $\frac{2}{3} + \frac{5}{6} + \frac{3}{8}$

10 Now complete the fraction calculations in **Q9**.

Factors

⊕ Find all the pairs of factors of a number
⊕ Find the highest common factor of two numbers

The **factors** of a number are all the numbers that divide exactly into that number.

1 is a factor of every number.

To find the factors of a number, start by trying to divide the number by 1, then by 2, then by 3, and so on.

The factors of 48 are: 1, 2, 3, 4, 6, 8, 12, 16, 24, 48
The **factor pairs** for 48 are: $1 \times 48, 2 \times 24, 3 \times 16, 4 \times 12, 6 \times 8$
The factors of 30 are: 1, 2, 3, 5, 6, 10, 15, 30
The factor pairs for 30 are: $1 \times 30, 2 \times 15, 3 \times 10, 5 \times 6$
The numbers that appear in both lists are called the **common factors** of 48 and 30. They are: 1, 2, 3, 6
The largest of the common factors is called the **highest common factor (HCF)**. So the HCF of 48 and 30 is 6.

Example Find a) the common factors of 60 and 24
b) the HCF (highest common factor) of 60 and 24.

Factors of 24: **1, 2, 3, 4, 6**, 8, **12**, 24

Factors of 60: **1, 2, 3, 4**, 5, **6**, 10, **12**, 15, 20, 30, 60

a) Common factors are 1, 2, 3, 4, 6, 12

b) HCF is 12

Exercise 8.8

❶ Find the missing factors:
a) factors of 15 are: 1, 3, ☐, 15
b) factors of 34: 1, ☐, ☐, 34
c) factors of 32: 1, 2, ☐, 8, 16, ☐
d) factors of 72: 1, 2, ☐, 4, ☐, 8, ☐, 12, ☐, ☐, 36, 72

2 List all the factor pairs of these numbers (3×4 is a factor pair of 12).
Now put your factors in order.

    a) 12   **b)** 20   **c)** 25   **d)** 36

3 Use your answers to **Q2** to find the common factors of:

    a) 20 and 12

    b) 20 and 25

    c) 12 and 36

4 Find the HCF of:

    a) 12 and 18

    b) 18 and 45

5 True or False?

    a) 9 is a factor of 27

    b) 6 is the largest number that divides exactly into 18 and 24

    c) 4 is a factor of 30 and 40

    d) 5 is the HCF of 20 and 25

    e) 6 is the HCF of 12 and 24

6 Copy and complete this HCF grid.

HCF	35	20
30		
19		
14		

7 By listing all the factors of each number, find the HCF of 16, 28 and 40.

Investigation

8 Find all the factor pairs of 4, 9, 16, 25, 36

What do you notice?

Can you find the next five numbers in this pattern?

Prime numbers

⊕ Recognise prime numbers up to 100

Prime numbers have exactly two **factors** .

The two factors are 1 and the number itself.

The only factor pair for 13 is 1×13.

So 13 has exactly two factors: 1 and 13.

13 is a prime number.

The prime numbers up to 100 are:

1	2	3	4	5	6	7	8	9	10
11	12	13	14	15	16	17	18	19	20
21	22	23	24	25	26	27	28	29	30
31	32	33	34	35	36	37	38	39	40
41	42	43	44	45	46	47	48	49	50
51	52	53	54	55	56	57	58	59	60
61	62	63	64	65	66	67	68	69	70
71	72	73	74	75	76	77	78	79	80
81	82	83	84	85	86	87	88	89	90
91	92	93	94	95	96	97	98	99	100

There are 25 prime numbers less than 100.

Note that 1 is not a prime number because it only has one factor.

Its factor pair is 1×1.

2 is the only even prime number.

Example   Which of these are prime numbers?

1   7   10   13   15

7 and 13 are prime numbers.

1 is not a prime number because it only has one factor.
10 is not a prime number because it has 4 factors (1, 2, 5 and 10).
15 is not a prime number because it has 4 factors (1, 3, 5 and 15).

Exercise 8.9

1 Are these prime numbers? Copy the number and write yes or no next to each one.

a) 7   b) 15   c) 33   d) 49

e) 51   f) 69   g) 31   h) 37

i) 81   j) 83   k) 93   l) 97

Look at lesson 1.3 Tests for divisibility.

2 Which number is not a prime number in these lists?

a) 1, 2, 3, 5, 7

b) 2, 3, 4, 5, 7

c) 5, 7, 9, 11, 13

d) 11, 13, 15, 17, 19

e) 31, 37, 41, 43, 45

3 List all the prime numbers up to 100.

4 Find the next prime number after each of these.

a) 16

b) 65

c) 44

d) 78

e) 80

f) 90

g) 52

h) 5

i) 32

j) 25

5 True or False?

a) 9 is a prime number.

b) 1 is a prime number.

c) Only one even number is a prime number.

d) Only prime numbers have exactly two factors.

e) There are ten prime numbers up to 100.

6 Roll two dice and decide which will be the tens digit and which will be the units digit of a 2-digit number.

If the number is a prime number score 5 points.

If the number is not a prime number lose 2 points.

Repeat ten times, recording your score each time.

What is your total score?

Repeat the activity and see if you can beat your score.

7 Use a set of 1–9 digit cards.

Shuffle them and deal out five cards.

Investigate how many different 2-digit prime numbers you can create with the five cards.

Repeat for a different set of cards.

Investigation

8 Apart from 2 and 5, all prime numbers have a units digit of 1, 3, 7 or 9. Carry out an investigation to see if this statement is true.

Reflection

◈ Know how to describe a reflection

◈ Recognise where a shape will be after reflection

Key words
reflect
mirror line
object
image
equivalent point
perpendicular bisector

- We can **reflect** shapes in a **mirror line**.

- The original shape is called the **object** and its reflection is called the **image**.

- Each point on the object has an **equivalent point** on the image. Equivalent points are the same distance from the mirror line. For example the point A is the same distance from the mirror line as A'.

These are equivalent points.

These distances are the same.

- The mirror line cuts the lines joining equivalent points on the object and image at right angles: it is called the **perpendicular bisector** of this line.

Example 1 Reflect this shape in the mirror line.

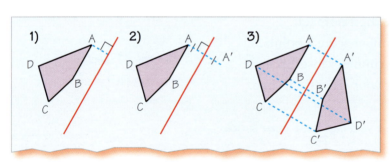

1) Label the vertices. Draw a straight line from A to cut the mirror line at 90°.

2) Measure the distance. Continue the line behind the mirror, measure the same distance and mark A'.

3) Repeat with the other vertices, and join up the points.

Example 2 Reflect ABCD in the y-axis.

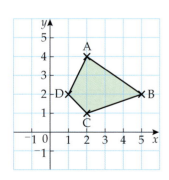
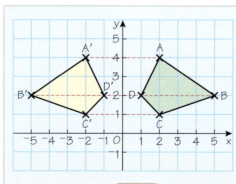

1) Find the distance between each point on the object and the y-axis. Continue the line through the y-axis and mark the same distance again. Mark the position of the new point.

2) Join up all the new points to make the image.

Exercise 9.1

1 Copy each diagram and reflect the shapes in the mirror lines.

Use a mirror to help you.

a) 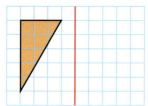   b) 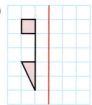   c) 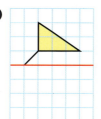   d)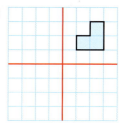

2 Copy each diagram and reflect the shapes in the mirror lines.

a)   b)   c)   d)

3 Copy each diagram and reflect the shapes first in the x-axis and then in the y-axis. For each diagram, write down the coordinates of the reflected shapes.

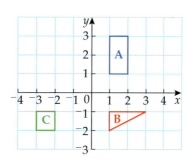

4 Reflect this quadrilateral in the line $x = 4$.

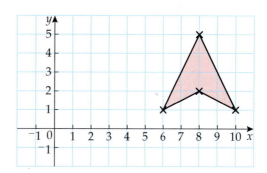

Investigation

5 Rangoli patterns are drawn as decorations during the festival of Divali. Here are two examples. Make up one of your own using either square or triangular dotty paper.

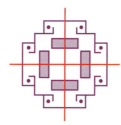   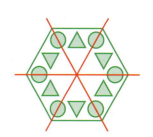

Rotation

- Know how to describe a rotation
- Recognise where a shape will be after rotation

Key words
rotation
angle of rotation
centre of rotation
clockwise
anticlockwise

A **rotation** is a turn. We describe a rotation by stating the **angle of rotation** (the angle through which a shape has moved) and where the **centre of rotation** is.

The centre of rotation can be inside, outside, or on the shape and is the only point that does not move during a rotation.

If a direction is not stated, it is assumed to be anticlockwise.

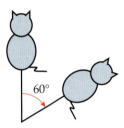

The cat has been rotated through 60° **clockwise**. To return it to the starting position, it must be rotated either through a further 300° clockwise or 60° **anticlockwise**.

Example 1 Rotate this shape through a half turn anticlockwise (180°) about the cross. Draw the image.

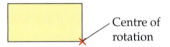

Centre of rotation

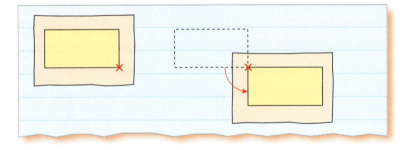

Place a piece of tracing paper over the shape and trace it. Put a pencil point over the cross so that the tracing paper doesn't slip and rotate the shape through 180°.

Example 2 Rotate this shape through a quarter of a turn (90°) clockwise about the cross. Draw the image.

×—— Centre of rotation

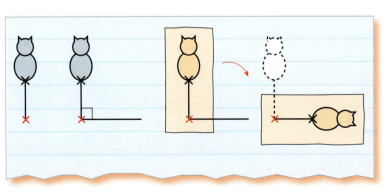

Join one part of the shape to the cross and draw an angle of 90°. Place a piece of tracing paper over the shape and trace it. Put a pencil point over the cross so that the tracing paper doesn't slip and rotate the shape through 90° clockwise.

Exercise 9.2

1 Copy these shapes and the centres of rotation. Use tracing paper to rotate them through 180°.

> Don't forget that if no direction is given, it is **anticlockwise**.

a)   b)   c) 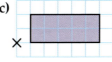   d) 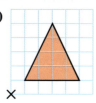   e)

2 Copy these shapes and the centres of rotation. Use tracing paper to rotate them through 90°.

a) 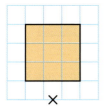   b) 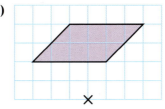   c)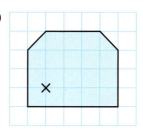

3 Rotate this shape through 180° clockwise about the point (0, 0). Write down the coordinates of the points on the image.

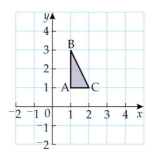

4 Use tracing paper to work out which is the centre of rotation for each pair of shapes.

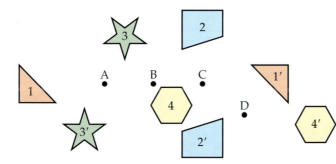

5 a) In **Q4**, the objects are marked with a number, such as 1, and the images are marked with a number and a dash, such as 1′. Describe what rotation is needed to move from each object to its corresponding image.

b) Describe what rotation is needed to move from each image back to the original object. (This is called the **inverse** transformation.)

Reflection and rotation symmetry

⊕ Recognise and explore reflection symmetry
⊕ Recognise and explore rotation symmetry

Key words
symmetrical
line of symmetry
mirror line
reflection symmetry
order of rotation symmetry

- If one half of a shape fits exactly onto the other half we say it is **symmetrical**. The line between the two halves is called the **line of symmetry** or **mirror line**. Shapes that are symmetrical about a line have **reflection symmetry**.

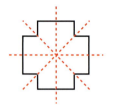

This shape has reflection symmetry. The lines of symmetry are marked in red.

- The **order of rotation symmetry** is the number of ways a shape can fit onto itself in a full 360° turn. All shapes have rotation symmetry of at least order one.

This shape has can fit onto itself in two different ways. It has rotation symmetry of order two.

- Regular polygons have the same number of lines of symmetry and order of rotation symmetry as sides.

Example   Look at the following shape.
Mark on all the lines of symmetry and state the order of rotation symmetry.

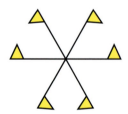

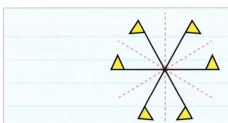

There are three lines of symmetry.

The shape fits onto itself in three different ways so it must have rotation symmetry of order three.

Use tracing paper to check for both reflection and rotation symmetry. For reflection symmetry, turn the paper **over** and for rotation symmetry turn the paper **round**.

Exercise 9.3

1 Each of these shapes is half of a symmetrical shape. Draw the complete shape.

> There may be more than one possible answer.

2 Write down the number of lines of symmetry and the order of rotation symmetry for each of these shapes:

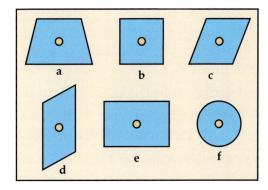

3 Add on one more square to make a shape with reflection symmetry.

   a) How many different solutions are there? Add on extra squares one at a time.

   b) How many squares would you need to make a shape with four lines of symmetry? Illustrate your answer.

4 Add on another triangle to each of the following to make shapes with rotation symmetry of order 3. Is there more than one solution for each shape?

   a)   **b)**   **c)**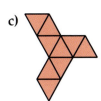

Investigation

5 Print out some of the characters from Wingdings or one of the other symbol fonts. For example: ❑ ⭢ ❋ ✿ ✎.   Find three characters which:

   a) have rotation symmetry of order one and no lines of symmetry

   b) have rotation symmetry of order two and no lines of symmetry

   c) have rotation symmetry of order one and one line of symmetry

   d) have rotation symmetry of order one and two lines of symmetry.

Translation

- Know how to describe a translation
- Recognise where a shape will be after translation

Key words
transformation
object
image
translation

A **transformation** moves a shape to a new position, from the **object** to the **image**.

Translation is the mathematical word for a slide.

A translation can move an object across, up or down.

To describe a translation, we have to give both the **distance** and the **direction**.

Here, a translation of **3 right**, **2 up** is shown.

We can tell how an object has been translated by looking at one point on the object and comparing it with the matching point on the image.

We count how far it has moved.

The rest of the shape will have moved in the same way.

We give the movement across (left or right) before the movement up or down.

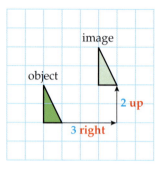

Example   Describe the translation necessary to move from:

a) the shaded object to A

b) the shaded object to B.

Choose a point on the object and compare it with the matching point on the image. Count how far it has moved.

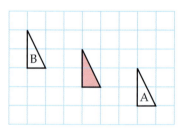

A: 3 right and 1 down

B: 3 left and 1 up

Exercise 9.4

1 This diagram shows a translation of Mr Poly's head.
  Describe the change of position of:

a) his eye

b) the end of his nose

c) his chin.

What do you notice?

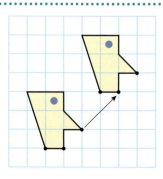

2 Make a copy of this diagram.
Translate the kite as instructed:

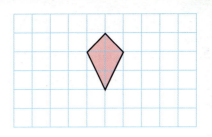

    a) 4 right and 3 down. Label the image A.

    b) 3 left and 2 up. Label the image B.

    c) 3 right and 2 up. Label the image C.

3 Describe the translation necessary to move from the shaded object to each image.

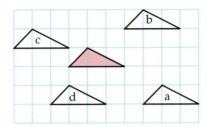

4 Make a copy of this diagram.

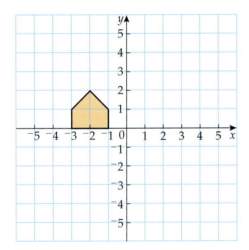

    a) Translate the shape 3 right and 2 up.
        Write down the coordinates of the image.

    b) Translate the shape 2 left and 3 up.
        Write down the coordinates of the image.

    c) Do you notice a connection between the coordinates of the object, the translation and
        the coordinates of the image?

5 Translate a shape four right and two up and draw the image. Then translate it three left
and five down. What transformation is needed to return it to the starting point?

Repeated transformations

⊕ Transform 2-D shapes using repeated reflections, rotations, and translations on paper and using ICT

- Reflection, rotation and translation are all **transformations** .

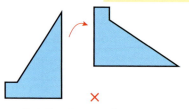
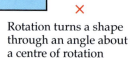
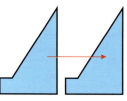

Reflection produces an image of a shape in a mirror line

Rotation turns a shape through an angle about a centre of rotation

A translation moves a shape up, down or across

- A transformation moves a shape to a new position, from the **object** to the **image** .
- The shape and size stay the same when an object is reflected, rotated or translated.

Example 1 Reflect the object P in the mirror line M$_1$. Call the image Q.
Reflect Q in the mirror line M$_2$.
Call the image R.

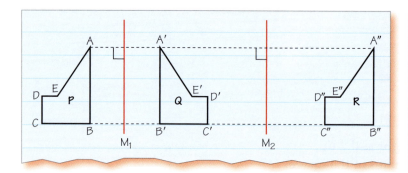

Label the vertices. Draw a straight line from A to cut the mirror line M$_1$ at 90°. Measure the distance. Continue the line behind the mirror and measure the same distance and mark A′. Repeat for the other vertices and join up the points. Repeat for Q in the mirror line M$_2$.

Example 2 Draw an equilateral triangle using a dynamic geometry program. By transforming the triangle and its images, fill the screen with tessellating equilateral triangles.

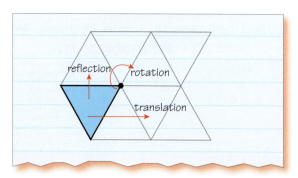

Decide where you will start and draw an equilateral triangle. Next, decide what transformations are necessary to fill the screen. You could fill the screen by repeatedly reflecting the triangle or by repeatedly rotating it. Alternatively, you could use a combination of transformations to fill the screen.

Exercise 9.5

1 Make a copy of this diagram.
Reflect the shape A in the x-axis and label the image B.
Reflect B in the y-axis and label the image C.

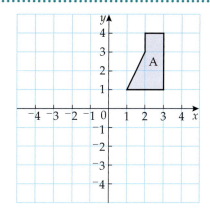

2 Make another copy of this diagram.
Rotate the shape A through 180° about the point (0, 0).
Label the image Z.
Look at your diagram for **Q1**.
What do you notice?

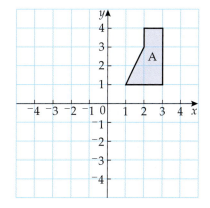

3 Copy this shape onto squared paper. Translate the shape 3 right and 4 down followed by 5 left and 3 up.
What single translation could you have used instead?

4 Make a copy of this diagram.
Rotate the shape 90° clockwise about (0, 0) followed by a rotation of 180° clockwise about (0, 0).
What single rotation could you have used instead?

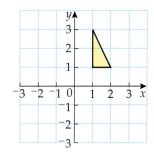

5 A **tessellation** is a tiling pattern with no gaps.
Tessellations can be made by repeatedly reflecting, rotating or translating shapes.
An example is shown.
Design a tessellation of your own, using a quadrilateral of your choice.

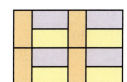

Drawing enlargements

⊕ Enlarge shapes given a centre of enlargement

Key words
enlargement
scale factor
centre of enlargement

Enlargement is a type of transformation. When a shape is enlarged, all the sides are made bigger by multiplying their lengths by a **scale factor** .

Every enlargement has a **centre of enlargement** .

Lines joining equivalent points on the object and image meet at the centre of enlargement.

Here, the triangle has been enlarged by a scale factor of 2. The enlargement is labelled A′B′C′.

The image is twice as far from the centre of enlargement (O) as the object.

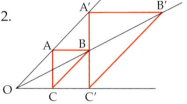

Example Enlarge rectangle ABCD by a scale factor of 3, using the point S as the centre of enlargement.

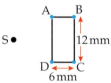

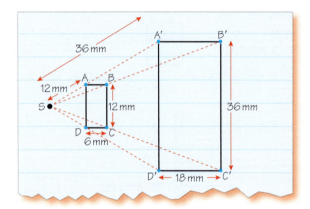

Draw the rectangle ABCD. Draw lines connecting the point S to the vertices A, B, C and D and extend them. Measure the lengths SA, SB, SC and SD. Multiply each measurement by three and then measure the new distances **from the point S**. Mark the points A′B′C′D′. Join A′B′C′D′.

Exercise 9.6

❶ Copy these shapes onto squared paper. Enlarge each shape by scale factor 2, using the point S as the centre of enlargement.

a)
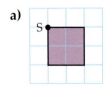

b)
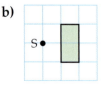

c)
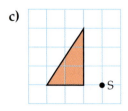

d)
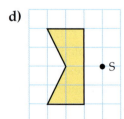

2 Copy these shapes onto squared paper. Enlarge each shape by scale factor 3, using the point S as the centre of enlargement.

a) 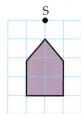 **b)**   **c)** 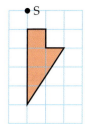 **d)**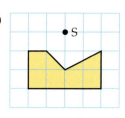

3 Trace each shape and enlarge it, from centre O, by the given scale factor.

a)

scale factor 4

b)

scale factor 3

c)

scale factor 2

4 a) Draw this shape on squared paper.

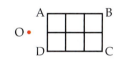

b) Enlarge the shape from centre O by the following scale factors:
   i) 1   **ii)** 2   **iii)** 3   **iv)** 4

c) Copy and complete this table of measurements:

Scale factor	1	2	3	4
Length AB	3			
Length BC	2			

d) Look at the lengths of the corresponding sides for the object ABCD and each of its enlarged images. What do you notice?

Investigation

5 Draw this shape with vertices at (0, 0) (0, 2) (2, 3) (2, 0) on a coordinate grid.

Enlarge the shape by scale factor 2, using (0, 0) as the centre of enlargement.

Write down the coordinates of the enlarged shape and compare them with the original coordinates.

What do you notice?

Investigate enlarging other shapes, using (0, 0) as the centre of enlargement.

You can choose which scale factor to use but make sure that one vertex of your shape is at (0, 0).

What do you notice about the coordinates of the vertices of the shape when it is enlarged?

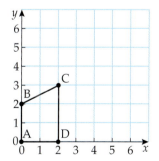

◈ Use letters to stand in for unknown numbers

◈ Express relationships between unknowns

Key words
unknown
relationship
algebraic expression

We can use letters to stand in for **unknown** numbers.

Sometimes we can express an unknown in terms of its **relationship** with another unknown.

For example, if p is an unknown number we can write:

three more than p	as	$p + 3$
two less than p	as	$p - 2$
five times p	as	$5p$
one third of p	as	$\dfrac{p}{3}$

We can also combine **algebraic expressions** . For example, $5p + p - 2 = 6p - 2$

Example   A plank of wood has length t cm.

Write an algebraic expression for a plank of wood whose length is:

a) four more than twice the length of t **b)** three less than half the length of t

c) three times one quarter of the length of t **d)** three times one more than the length of t.

a) Four more than twice the length.

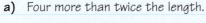

$2t + 4$ Twice the length is 2t. To find four more than this add 4.

b) Three less than half the length.

$\dfrac{t}{2} - 3$

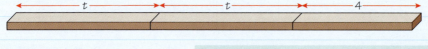

To find half the length divide by 2. To find three less than this subtract 3.

c) Three times one quarter of the length.

$\dfrac{3t}{4}$

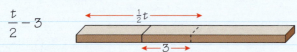

One quarter of the length is $\dfrac{t}{4}$, to find three times this, multiply by 3.

d) One more than the length, times three.

$3(t + 1)$ One more than the length is $t + 1$. To find three times this multiply by 3.

Exercise 10.1

1 Georgina receives y pounds pocket money.
Write down an algebraic expression for the amount each of the following people receive:
   a) Amanda receives £2 more than Georgina.
   b) Naomi receives £3 less than Georgina.
   c) Maggie receives three times as much as Georgina.
   d) Craig receives half as much as Georgina.
   e) Harpreet receives twice as much as Amanda.
   f) Lyndon receives a half as much as Naomi.
   g) David receives £4 more than Maggie.
   h) Catherine receives three times more than Craig.

2 Look at your answers to **Q1**.
   a) If Georgina receives £5 pocket money, write down how much Amanda, Naomi, Maggie, Craig, Harpreet, Lyndon, David and Catherine receive.
   b) Who receives the most pocket money?
   c) Who receives the least pocket money?

3 A length of string is x cm long.
Write down the length of a piece of string that is:

   a) 2 cm shorter
   b) 13 cm longer
   c) five times the length
   d) one sixth of the length
   e) four less than double the length
   f) three more than half the length
   g) two more than the length, times eight
   h) three times half the length.

> Don't forget to use brackets.

4 Sasha and Jordan are using ribbon to wrap presents.
   a) Sasha cuts a piece of ribbon y cm long. Jordan cuts a piece 2 cm longer, then cuts this into three equal pieces. Write an expression for the length of each piece.
   b) Jordan cuts a piece h cm long. Sasha cuts a piece 3 cm shorter, then cuts this into four equal pieces. Write an expression for the length of each piece.

5 Lord Number collects china vegetables.
At the collectors' annual meeting he writes algebraic expressions to show how the number of china vegetables in other Lords' collections compare to his. His expressions are:

Lord Number	w	China vegetables
Lord Breakalotte	$w - 15$	China vegetables
Lord Losealotte	$\dfrac{w}{5}$	China vegetables
Lord Lostmore	$\dfrac{3w}{10}$	China vegetables
Lord Keepalotte	$5w + 4$	China vegetables

   a) For each Lord, describe in words how the number of china vegetables compares to Lord Number's collection. For example: Lord Breakalotte has fifteen less than Lord Number.
   b) Who has the most china vegetables?
   c) If Lord Number has 250 china vegetables, how many do each of the other Lords have?

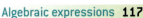

Why simplify?

◈ Solve equations

◈ Simplify expressions to make an equation easier to solve

You can **solve** equations using inverse operations or the balancing method.

For example, to solve $3x - 5 = 1$

Using **inverse operations**:

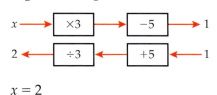

$x = 2$

Using the **balancing method**:

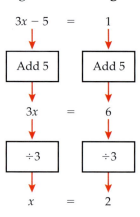

Before you attempt to solve an equation, **simplify** both sides of the equation as much as possible by **collecting like terms** .

Example   Solve the equation $37 = 6x + 2x - 3$

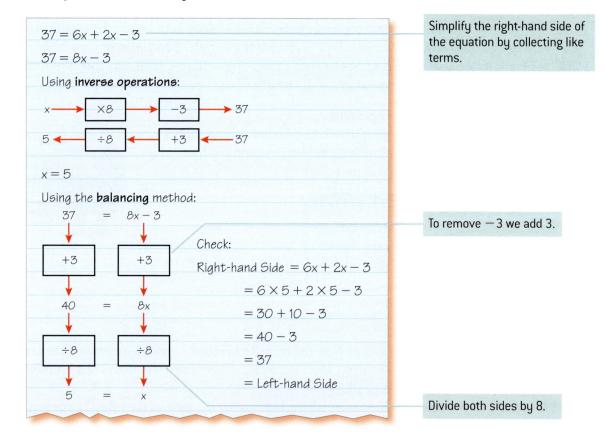

$37 = 6x + 2x - 3$

$37 = 8x - 3$

Using **inverse operations**:

$x = 5$

Using the **balancing** method:

$37 = 8x - 3$

$40 = 8x$

$5 = x$

Simplify the right-hand side of the equation by collecting like terms.

To remove -3 we add 3.

Check:

Right-hand Side $= 6x + 2x - 3$

$= 6 \times 5 + 2 \times 5 - 3$

$= 30 + 10 - 3$

$= 40 - 3$

$= 37$

$=$ Left-hand Side

Divide both sides by 8.

Exercise 10.2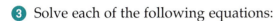

1 Solve each of the following equations. Don't forget to check your answers.

a) $5y + 3y = 48$ b) $10t - 1t = 27$ c) $8q + 12q = 100$

d) $14m - 11m = 24$ e) $8r + r = 99$ f) $49 = 2w + 5w$

g) $48 = 7b + 5b$ h) $63 = 12y - 3y$ i) $100 = 10m + 15m$

> Simplify as much as possible first.

2 Solve each of the following equations:

a) $3a + 2 + 5a = 26$ b) $3b - 2 + 5b = 54$ c) $2t + t - 5 = 19$

d) $10d - 3d + 3 = 80$ e) $4e + 2 - e = 32$ f) $4f - 7 - 2f = 1$

g) $68 = 9g + 5 - 2g$ h) $3 = 2h - h + 1$ i) $0 = 10i - 1 - 9i$

3 Solve each of the following equations:

a) $2x + 4 + 3x + 2 = 16$ b) $5x - 2x + 2 - 1 = 37$

c) $7x - 3 + 4 + 3x = 41$ d) $8x + 2 - 7x - 2 = 6$

e) $12 = 12x - 5 - 6x + 11$ f) $61 = 7x + 2x + 3x + 1$

g) $40 = 15x + 5 - 15 - 10x$ h) $360 = 10x + 12x - 100 + 20$

4 a) Write down an algebraic expression for the sum of the angles in the diagram.

b) Simplify your answer to part **a)**.

c) The angles on a straight line add up to 180°.
   Use this fact to copy and complete this equation:
   $$\ldots\ldots = 180°$$

d) Solve your equation from part **c)** to find the value of x.

e) What are the sizes of the three angles on the line?

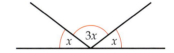

5 The area of each pink rectangle is $4x$ and the area of each green rectangle is $3x$.

a) Write down and simplify an algebraic expression for the total area of:
      i) the pink rectangles   ii) the green rectangles.

b) Add your answers from part **a)** to write down an expression for the total area of the shape. Simplify your expression.

c) The area of the shape is 120 cm². Form an equation in order to find x.

d) Solve your equation.

e) What is the area of a pink rectangle?

f) What is the area of a green rectangle?

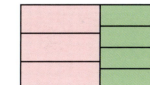

6 Three friends do a sponsored walk for charity. Alix raises y pounds.

a) Ben raises four times as much as Alix. Write an expression for the amount he raises.

b) Cassie raises £7 less than Alix. Write an expression for the amount she raises.

c) Write down and simplify an expression for the total amount raised by the three friends.

d) The total raised is £53. Write down and solve an equation to find the value of y.

e) How much did each person raise?

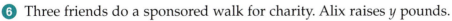

Solving equations involving divisors

⊕ Solve equations where the unknown is divided by a number

You can **solve** equations where the unknown is divided by a number using inverse operations or the balancing method.

For example, $\dfrac{n}{4} = 9$

Using **inverse operations**:

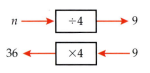

$n = 36$

Using the **balancing** method:

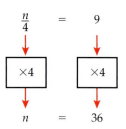

The unknown value 'n' has been divided by 4. The inverse of dividing by 4 is multiplying by 4.

Check:
Left-hand Side $= \dfrac{n}{4}$

$= \dfrac{36}{4}$

$= 9$

$=$ Right-hand Side

Example   Find the value of a when: $8 = \dfrac{4a}{5}$

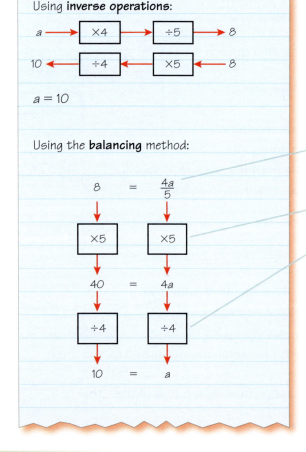

a has been multiplied by 4 and then divided by 5. We do the inverse operations in the reverse order.

The inverse of dividing by 5 is multiplying by 5.

The inverse of multiplying by 4 is dividing by 4.

Check:
Right-hand Side $= \dfrac{4a}{5}$

$= \dfrac{4 \times 10}{5}$

$= \dfrac{40}{5}$

$= 8$

$=$ Left-hand Side

Exercise 10.3

1 Solve each of the following equations. Don't forget to check your answers.

a) $\dfrac{n}{5} = 12$ b) $\dfrac{b}{3} = 20$ c) $8 = \dfrac{p}{4}$ d) $19 = \dfrac{q}{10}$

See the Explanation box on page 120.

e) $\dfrac{z}{2} = 12$ f) $11 = \dfrac{y}{5}$ g) $\dfrac{m}{6} = 7$ h) $\dfrac{r}{100} = 2$

2 Solve the following equations:

a) $\dfrac{2v}{5} = 4$ b) $\dfrac{7n}{2} = 14$ c) $8 = \dfrac{4k}{5}$ d) $\dfrac{2x}{3} = 4$

See the Example on page 120.

e) $5 = \dfrac{5p}{2}$ f) $30 = \dfrac{10t}{2}$ g) $\dfrac{3m}{10} = 9$ h) $1000 = \dfrac{100m}{3}$

3 A toyshop owner divides a large box of marbles into 20 separate bags.

a) There are m marbles in the box.
Copy and complete the algebraic expression for the number of marbles in one bag:

b) Each bag contains 5 marbles. Copy and complete this algebraic equation:

c) Solve your answer to part **b)** to work out the number of marbles in the box.

4 Five friends share a packet of n sweets equally between them.

a) Write an algebraic expression for the number of sweets each person has.

b) Each person has 4 sweets. Write an equation using your expression from part **a)**.

c) Solve your equation to find the number of sweets in the packet.

5 Football fans arrive at a match by coach. x fans come in 25 coaches.

a) Write an expression for the number of fans in a coach.

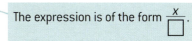
The expression is of the form $\dfrac{x}{\square}$.

b) Each coach carries 40 people. Write an equation and solve it to find the number of fans at the match.

c) How many coaches would be needed for 350 fans?

6 You can use the following formula to estimate the amount of income tax to pay on your earnings.

$$\text{income tax} = \frac{2 \times earnings}{10}$$

a) Calculate how much income tax you pay if your earnings are:

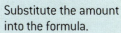
Substitute the amount into the formula.

  i) £25 000   **ii)** £10 000   **iii)** £12 500

b) Mr Morgan estimates that he will pay £300 income tax. What are his earnings?

c) Mrs Attwood estimates that she will pay £50 income tax. What are her earnings?

7 a) For the following equation, copy and complete the function machine below:

$\dfrac{x}{3} + 1 = 5$

b) What is the value of x?

8 Use your method from **Q7** to solve the following equations:

a) $\dfrac{x}{7} + 2 = 10$ b) $\dfrac{x}{3} + 4 = 6$ c) $11 = \dfrac{x}{12} + 8$

d) $4 = \dfrac{x}{5} - 6$ e) $2 = \dfrac{x}{2} - 3$ f) $\dfrac{x}{11} - 1 = 10$

◈ Form and solve equations for real-life situations

We often use **algebra** to solve problems.

If we pay 90p for 3 packets of crisps, we can easily work out that each packet costs 30p. We can write this **algebraically** as $3x = 90$

and solve it using **inverse operations**: or the **balancing** method:

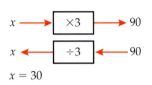

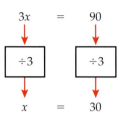

Example Find the size of the angles in this triangle:

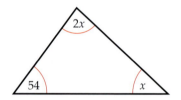

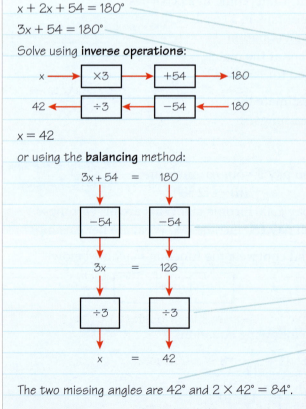

$x + 2x + 54 = 180°$

$3x + 54 = 180°$

Solve using **inverse operations**:

$x = 42$

or using the **balancing** method:

The angles in a triangle add up to 180°. Write an equation to show this.

Simplify the equation.

Subtract 54 from both sides.

Divide both sides by 3.

Make sure you answer the question you were asked.

The two missing angles are 42° and 2 × 42° = 84°.

Check: 42° + 84° + 54° = 180° ✓

Exercise 10.4

1 Find the size of the missing angles in the following triangles:

a)

$3x$
20
x

b)

$2y$
y

c)

This is an equilateral triangle.

z

2 The perimeter of a rectangular swimming pool is 80 m. The length of the pool is 30 m.

The perimeter is the distance all the way round the shape.

a) Sketch a plan of the swimming pool and label the lengths 30 m. Label the widths y.

b) Copy and complete this equation for the perimeter of the swimming pool:
$30 + 30 + y + y = \square$

c) Simplify your equation from part **b)**. Solve it to find the width of the pool.

3 **a)** A rectangular flower garden has perimeter 60 m. The width of the garden is 5 m. Write an equation and solve it to find the length of the flower garden.

b) A large square fish pond has perimeter 80 m. What is the length of each side?

Form an equation and solve it.

4 Three consecutive numbers add up to 21.

Consecutive numbers are ones which are next to each other, e.g. 3, 4 and 5.

a) Let n stand in for the smallest number. Write algebraic expressions for the next two numbers.

b) Form an algebraic equation and solve it to find n.

c) What are the three numbers?

5 The cost of this shopping bill is £5.25: The cheese costs £1.20 and the packets of biscuits are 50p each.

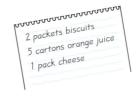

2 packets biscuits
5 cartons orange juice
1 pack cheese

a) Write an equation for this information.

b) Solve your equation to find the cost of one carton of orange juice.

6 In class 8L, the heights of five people were measured. The mean was calculated to be 1.5 m.

To find the mean you add all the heights together and divide by the number of people.

Four of the heights were: 1.4 m, 1.2 m, 1.5 m, 1.8 m

a) Write an equation for the mean height.

b) Solve your equation to find the height of the fifth person.

Investigation

7 Two whole numbers add together to give 10. They multiply together to give 21. Let one of the numbers be a and the other b.

a) Write down two algebraic equations to show this information.

b) Find the values of a and b.   Try different values of a.

c) Why couldn't you solve the equation using your usual method?

⊕ Construct formulae in words

⊕ Substitute whole numbers into formulae expressed in words

A **formula** expresses a relationship between two or more **variables**. Variables can stand in for different values.

If you have to do lots of similar calculations, it may be easier to construct a formula first.

For example, a farmer sells eggs in boxes of 6.

He can construct a formula connecting the number of eggs and the number of boxes.

$$\text{Number of boxes} = \frac{\text{number of eggs}}{6}$$

Shop A orders 300 eggs.

Shop B orders 360 eggs.

He can use his formula to calculate how many boxes he needs for each shop:

$$\text{Shop A: Number of boxes} = \frac{300}{6} = 50$$

$$\text{Shop B: Number of boxes} = \frac{360}{6} = 60$$

Example To make pastry for pies, a baker uses twice as much flour as fat.

     a) Write a formula connecting the amounts of fat and flour.

     b) Use your formula to calculate how much flour is needed to make pastry with:

        i) 250 g fat   **ii)** 1 kg fat.

a) amount of flour = 2 × amount of fat	If you know the amount of fat, you double it to find the amount of flour.
b) i) amount of flour = 2 × 250 g = 500 g	The fat is measured in grams, so the flour is measured in grams.
ii) amount of flour = 2 × 1 kg = 2 kg	The fat is measured in kg, so the flour is measured in kg.

Exercise 10.5

1 A sweet shop manager packs chocolates in boxes of 9.

    a) Write a formula connecting the number of boxes and the number of chocolates.   Number of boxes = ...

    b) How many chocolates does the manager need for each of these customer orders?

        i) Mrs Smith: 11 boxes   **ii)** Mr Patel: 17 boxes   **iii)** Miss Chung: 45 boxes.

    c) A customer orders six boxes of chocolates. How many chocolates does the manager need to fill these?

2 The school caretaker sets out rows of chairs for a parents' meeting.
There are 15 chairs in each row.

   a) Write a formula connecting the number of rows to the number of chairs.

> Number of rows = ...

   b) Use your formula to calculate how many rows of chairs are needed for:

      i) 150 parents   **ii)** 225 parents   **iii)** 300 parents.

3 To make Swiss roll, a baker uses 30 g of flour and 25 g of sugar for every egg used.

   a) Write a formula connecting the amount of flour and number of eggs used.

   b) Use your formula to calculate the amount of flour needed for a Swiss roll made with:

      i) 3 eggs   **ii)** 5 eggs.

   c) Write a formula connecting the amount of sugar and number of eggs used.

   d) Use your formula from part **c)** to calculate the amount of sugar needed for a Swiss roll made with:

      i) 2 eggs   **ii)** 6 eggs.

   e) Copy and complete this recipe:

Swiss roll
7 eggs
☐ g flour
☐ g sugar

> Use your formulae to find the missing amounts.

4 **a)** Write a formula for the perimeter of this rectangle. Simplify your formula as much as possible.

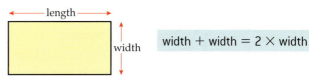

length

width

> width + width = 2 × width

   b) A hockey pitch is 90 m by 60 m. Work out the length of the white line all around the outside of the pitch.
For a hockey match, a rope fence is put around the pitch to keep the spectators back. The fence is 3 metres away from the edge of the pitch.

   c) Draw a diagram of the hockey pitch and the fence around it. Mark on all the measurements.

   d) How long is the rope fence?

5 Petrol costs 75p per litre.

   a) Write a formula connecting the cost of petrol and the number of litres used.

   b) A car travels 10 km on 1 litre of petrol.
Write a formula connecting the number of litres used and km travelled.

   c) Substitute your formula from part **b)** into your formula from part **a)**. This gives you a new formula connecting the cost of petrol with km travelled.

   d) Calculate the cost of petrol for these journeys:

      i) 50 km   **ii)** 75 km   **iii)** 500 km.
Convert your answers to pounds.

   e) Change your formula from part **c)** so that it calculates the cost of petrol in pounds. Use it to calculate the cost of petrol for a 450 km journey.

10.6 Constructing formulae

◈ Derive simple formulae

◈ Use formulae to solve problems

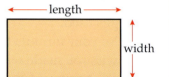

A **formula** expresses a relationship between two or more **variables** .
A variable is an unknown that can have more than one value.
Using a formula can help us to **solve** problems quickly.

We can use letters to stand in for variables in formulae.
For example, the formula for the area of a rectangle is:

Area = length × width

We can write this as:

$A = lw$

> You must explain what the letters mean!

where l = length and w = width

We can find the area of any rectangle by **substituting** values of l and w into the formula.

When constructing formulae, make sure your information is in the correct units.

length / width

Example
A cake recipe uses 2 oz of flour for every egg used.
Let e represent the number of eggs and f represent the amount of flour (in ounces).

a) Write down a formula connecting f to e.

b) Calculate the amount of flour needed for 27 eggs.

c) If 110 oz of flour is used, how many eggs are needed?

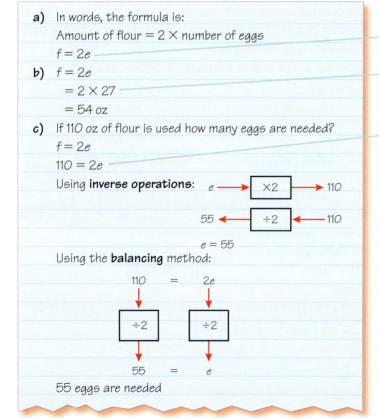

a) In words, the formula is:

Amount of flour = 2 × number of eggs

$f = 2e$

b) $f = 2e$

$= 2 × 27$

$= 54$ oz

c) If 110 oz of flour is used how many eggs are needed?

$f = 2e$

$110 = 2e$

Using **inverse operations**:
$e \longrightarrow \boxed{×2} \longrightarrow 110$
$55 \longleftarrow \boxed{÷2} \longleftarrow 110$
$e = 55$

Using the **balancing** method:
$110 = 2e$
$\boxed{÷2} \quad \boxed{÷2}$
$55 = e$

55 eggs are needed

Replace 'amount of flour' with f and 'number of eggs' with e.

Substitute the value $e = 27$ into the formula.

Substitute the value $f = 110$ into the formula.

Solve the equation to find the value of e.

Exercise 10.6

1 In a phone box calls cost 10p per minute.

> In your formula, use *c* for the cost and *n* for the number of minutes.

    a) Write a formula connecting the cost of the call, *c* to the length of the call in minutes, *n*.

    b) Use your formula to calculate the cost (in pence) of a call lasting:

        i) 20 minutes   **ii)** 5 minutes

2 1 litre of paint will cover 6 m² of wall.

    a) How many litres of paint would you need to cover these areas?

        i) 12 m²   **ii)** 36 m²   **iii)** 60 m²

    b) Write a formula connecting the area of the wall, *n* to the litres of paint needed, *k*.

    c) Use your formula to answer the questions in part **a)**. If your formula is correct you should get the same answers.

3 Alexander is two years older than his sister Emily. He always receives £2 more pocket money than Emily.

> You could use *A* for the amount of pocket money Alexander receives and *E* for the amount Emily receives.

    a) Write down a formula connecting the amount of pocket money Alexander and Emily receive.

    b) Use your formula to calculate the amount of pocket money Alexander receives if Emily receives:

> Be careful with the units!

        i) £1   **ii)** £1.50   **iii)** £3   **iv)** 90p

    c) Use your formula to calculate how much Emily receives when Alexander receives:

        i) £6   **ii)** £8   **iii)** £7.50

4 A Taxi company charges the following:

> ## ROVERS RIDES
> £1 per mile + £2 basic charge.

    a) How much would a journey of 3 miles cost?

    b) Write a formula in words to calculate the cost of a journey with Rovers Rides.

    c) Write your formula using algebra. Use *R* for the cost of the journey and *m* for the distance in miles.

    d) Use your formula to calculate the cost of these journeys:

        i) 20 miles   **ii)** 13 miles   **iii)** 70 miles

    e) Work out how long the journey was if it cost:

        i) £32   **ii)** £45   **iii)** £2.50

Investigation

5 Another Taxi company has the following charges:

> ## CANDICE CARS
> 50p per mile + £3 basic charge.

    a) How much would a journey of 3 miles cost?

    b) Write a formula for calculating the cost of a journey with Candice Cars.

> Try substituting different distances into your formulae.

    c) For which distance do the two taxi companies charge the same amount?

Two-way tables

⊕ Design, use and interpret two-way tables

A **two-way table** shows information sorted into different categories.

Example 1 The two-way table shows the results for England of the One Day International cricket matches between England and Australia from 1990–2002.

Result / Where played	Won	Lost
England	3	6
Australia	4	12

a) How many matches were played in total?
b) What percentage of the games were played in Australia?
c) What percentage of the games did England lose in total?

a) $3 + 6 + 4 + 12 = 25$ matches

b) 16 games. $\frac{16}{25} \times 100 = 64\%$

c) 18 games. $\frac{18}{25} \times 100 = 72\%$

Example 2 In a class of 20 pupils, 10 children wear glasses. There are 9 girls in the class.

	Girl	Boy
Glasses	4	
No glasses		

a) Copy and complete the two-way table to show this information.

b) What fraction of the class are boys who do not wear glasses?

a)

	Girl	Boy
Glasses	4	6
No glasses	5	5

This makes 10 children wearing glasses.

This makes 9 girls.

b) $\frac{5}{20}$ or $\frac{1}{4}$

5 boys out of a total of 20 children do not wear glasses.

Exercise 11.1

1. The table shows the type and outcome of the football matches played by Burnley football club in the 2001–2 season.

	Home	Away
Win	10	5
Draw	4	6
Lose	9	12

a) How many games were played in total?
b) How many games did Burnley win in total?
c) How many games were lost at home?
d) How many games were not drawn?

2 In a class of 30 pupils, 20 pupils have packed lunch. The remainder have school dinners. There are 15 boys in the class.
Complete the table to show this information.

Gender Type of lunch	Boys	Girls
Packed lunch	7	
School dinner		

3 A group of 10 people each buy a drink and a packet of crisps.

Use this information to complete the table.

6 Colas
7 packets plain crisps
3 packets cheese and onion crisps

	Plain crisps	Cheese and onion crisps
Cola	4	
Orange juice		

a) How many people chose orange juice?
b) What percentage of people chose plain crisps?
c) How many people chose orange juice and cheese and onion crisps? What is this as a percentage?

4 This table shows the distance in miles between different towns and cities

London			
57	Oxford		
80	150	Dover	
120	74	210	Bristol

a) What is the distance from Oxford to Bristol?
b) Which two towns are closest together?
c) Which two towns are furthest apart?
d) A coach travels from Oxford to Dover and back again. How far is this journey in total?
e) A lorry driver travels from London to Oxford and then to Bristol. She returns to London from Bristol. How many miles has she driven?

5 The cost of ten-pin bowling depends on the time of day.

	Daytime	Evenings during the week	Weekends
Adults	£1.00	£2.50	£3.00
Children	£0.50	£2.00	£2.50

a) What is the total cost for one adult and three children during the daytime?
b) How much more would it cost the same group to go at the weekends?
c) A group of children go bowling during the day. It costs £7.50 in total. How many children went bowling?
d) A group of four people pay £10 to go bowling. When did they go and were they adults or children?   There are two possible answers.

Investigation

6 'There are more left-handed boys than there are left-handed girls'

Design a two-way table to show gender against writing hand.
Complete the table by asking pupils in your maths group.
Do you agree with the statement above?

Averages

⊕ Calculate the mode, median, mean and range where appropriate

Key words
mode
frequency
median
mean
range

The three most widely used averages are:

The **mode** : item of data with the highest **frequency** (that is, the value that occurs most often).

The **median** : middle value once the data is ordered. If there is an even number of values it is the mean of the middle two values.

The **mean** : sum of the data divided by the number of items in the data set.

The **range** : the largest value in the data set minus the smallest. This shows the spread of the data and is not an average.

Example   Meena visits different shops looking at the cost of a can of lemonade.

25p   45p   30p   25p   40p   35p   42p   30p   27p   34p   30p   45p   34p

Calculate the mode, median, mean and range for her information.

25p 25p 27p 30p 30p 30p

34p 34p 35p 40p 42p 45p 45p

The mode is 30p

The median is 34p

The mean $= \dfrac{25 + 25 + 27 + 30 + 30 + 30 + 34 + 34 + 35 + 40 + 42 + 45 + 45}{12}$

$= \dfrac{442}{13}$

$\dfrac{442}{13} = 34p$

The range is 45p − 25p = 20p

Write the data in order.

There are more values of 30p than any other.

The median is the middle value of an ordered set of data.

Exercise 11.2

① A group of children each buy 100 g of pick and mix. Here are the number of sweets each child buys: 14   15   16   18   18   20   20   20   21

Calculate the mode, median, mean and the range for the number of sweets in 100 g of pick and mix.

② The number of letters Sam receives each day are shown in this table:

Day of the week	Saturday	Sunday	Monday	Tuesday	Wednesday	Thursday	Friday
Number of letters	3	0	1	2	0	1	0

Calculate the mode, median, mean and the range for the number of letters he receives.

Remember to put the information in order.

3 This is how many apples were counted in a number of bags:

6   6   7   7   7   7   8   8   8   10   10   12

Calculate the three averages and decide which average was probably used.

4 Three form groups raise some money from a sponsored walk.

Form group	Amount raised	Number of pupils
8A	£114.75	27
8B	£107	25
8C	£125.70	30

a) Which form group raised the most money in total?
b) Which form group raised the most money per pupil?

> Divide the amount raised by the number of pupils for each form group.

5 Helen receives £3.75 each week for spending money. She tells her parents that this is 'below average'. She asked eight of her friends how much spending money they receive each week. Here are their replies.

£3   £3   £3.50   £3.50   £3.50   £4   £5.50   £6

a) Which average would Helen use to prove to her parents that her spending money is below average?
b) Which average would her parents use to prove that her spending money is above average?

> Calculate each of the three averages.

6 The Jones family take a canal barge holiday. To complete the Midlands loop, the barge needs to cover a mean distance of 25 km per day. Mr Jones records the distance travelled each day.

Day	Saturday	Sunday	Monday	Tuesday	Wednesday	Thursday	Friday
Distance (km)	28	21	25	26	19	29	

The most the barge can travel is 30 km per day.
Can they complete the loop on Friday? Explain your answer.

> Find the **total** they need to cover in 7 days.

Investigation

7 Requires today's newspaper.

Use the weather information in a newspaper to calculate the 'average' temperature in the United Kingdom yesterday.

Should you calculate all three averages?

> Do you need to use all of the information provided?

11.3 Frequency tables and calculating the mean

⊕ Record data in frequency tables
⊕ Use frequency tables to calculate the mean

Key words
event
frequency
frequency table
mean

The word **event** is used to describe something in particular that happens. For example, an event is that I roll a 3 on a dice.

We use the word **frequency** to describe the number of times an event happens.

A **frequency table** is a way of sorting this data into groups. To find the **mean**, instead of adding up all the bits of data separately we can find the totals for each group.

Example 1   A teacher keeps a record of the number of pupils attending netball practice each week. Draw a frequency table for this data.

12 13 12 10 12 13 12 13 12 12 10 13 12 11

Number of pupils	Tally	Frequency
10				2
11			1
12								7
13						4

Example 2   Eight people visit the cinema. The three adults pay £6 each, four children each pay £4, and one senior pays £2. Make a frequency table and use it to calculate the mean cost per person.

	Cost per person (£)	Number of people (frequency)	Frequency × cost (£)
Adult	6	3	18 (3 × 6)
Child	4	4	16 (4 × 4)
Senior	2	1	2 (1 × 2)
		Total number of people = 8	Total cost = £36

The mean cost = total cost ÷ number of people = £36 ÷ 8 = £4.50

Exercise 11.3

1 The highest breaks in snooker from 1988–2000 at the World Snooker Championships were:

140   141   140   140   147   144   143   147   144   147   143   142   143

Make a frequency table for this data.

2 Jane makes a frequency table to show the different coins in her money box.

Copy and complete the table and find the total amount she has in her money box.

Coins	Frequency	Amount
1p	16	16p (1p × 16)
2p	11	
5p	8	
10p	6	
		Total amount =

3 A P.E. teacher records the number of goals scored by the school football team in 20 matches.

3 1 2 0 4 2 2 1 3 0
2 4 0 2 1 0 1 1 1 0

a) Copy and complete the frequency table for this data.
b) Work out the mean goal score per match.

Number of goals	Tally	Frequency
0		
1		
2		
3		
4		

4 A group of people travel on the Underground. Copy and complete the table and calculate the mean cost per person for travelling on the Underground.

	Cost	Frequency	Frequency × cost
Adult	£3	1	
Child	£2	6	
Senior	£1. 50	2	
		Total frequency =	Total cost =

Investigation

5 This question requires a dice.
Estimate how many throws of a dice you need, on average, to get a score of six or more. Throw a dice, and count the throws needed to reach a total score of six or more. Repeat the experiment 20 times recording the number of throws in a table like the one shown. Calculate the mean number of throws – how close was your estimate?

Number of throws	Tally	Frequency	Number of throws × frequency
1			
2			
3			
4			
5			
6			
		Total frequency =	Total number of throws =

Comparing two distributions

⊕ Compare two distributions using the range and one or more of the mode, median and mean

A **distribution** tells you the frequency of the values in a set of data. You can compare two distributions using **statistics** – the mode, median, mean and range.

The mode: the value with the highest frequency.

The median: middle value once the data is ordered.

The mean: sum of the data divided by the number of values in the data set.

The range: the largest value in the data set minus the smallest.

You will not always need to calculate all three averages.

Example   Greta has the choice of two different taxi firms. The table shows how many minutes she waits for a taxi to arrive.

Ace cabs	6	6	6	7	7	8	8	10	10	12
Kwik kars	3	4	4	5	6	10	11	11	11	15

Which taxi company should she use so that on average she waits for less time?

	Mean	Median	Mode	Range
Ace cabs	8	7.5	6	6
Kwik kars	8	8	11	12

Greta should choose 'Ace cabs' as both the median and mode are lower than Kwik kars. Also a large range for Kwik kars shows that sometimes Greta must wait a long time for a cab.

Exercise 11.4

1. Paul has the choice of two bus routes to get to school.

Route A	1	2	8	9	9	12	15
Route B	6	6	6	8	8	9	13

Copy and complete the table. Which bus route 'on average' has the smallest waiting time?

	Mean	Median	Mode
Route A			
Route B			

2 The school nurse measures the heights in centimetres of a group of boys and girls.

Girls' heights	135	137	141	141	146	152
Boys' heights	136	137	138	140	143	143

Calculate the three averages for both boys and girls.
Which averages show that the girls are taller than the boys?

Draw a table similar to the one in **Q1**.

3 Five pupils from Years 8 and 9 were asked how many visits to the library they made in a four-week period.

	Pupil 1	Pupil 2	Pupil 3	Pupil 4	Pupil 5
Year 8 pupils	2	7	1	6	4
Year 9 pupils	0	9	3	8	4

a) Which averages can be found?
b) Find these averages for each year group.

4 The results of six competitions where two archers fire 10 arrows at a target and the number of 'hits' are counted are shown in this table.

Competition number	1	2	3	4	5	6
Archer A	0	10	9	0	8	0
Archer B	5	1	4	3	6	5

Calculate the mode, mean and range of the scores of each archer.
Only one archer can enter the next competition. Explain which one you would choose.

5 The results of eight games between Manchester United and Arsenal are shown.

Manchester United	0	1	2	0	2	0	1	0
Arsenal	0	1	2	2	0	1	3	4

a) Calculate the three averages and the range for each team.
b) Write a sentence to compare the performance of the two teams using your answers from part **a)**.

6 The amount of rain each month in millimetres for two Scottish towns are shown.

	Mean	Median	Range
Fort William	165	151	135
Perth	62	66	39

a) Which town had the most rain? Explain how you know.
b) Fort William has a very big range. What does this tell us about the weather there?

11.5 Grouping data

Key words
class interval
modal class
modal group

- Construct frequency tables for data
- Find the range of a set of data and put data into equal groups
- Find the modal class for a set of grouped data

If data has been measured – such as time or length – or has a large range (spread) then it is often grouped. These groups are called **class intervals** . The class interval with the highest frequency is the **modal class** or **modal group** .

Example   Ian and Lynn are comparing numbers of birds in their garden at different times of year. Ian counted the birds he could see at 9 am for 30 days in June.

9	7	0	9	8	3	12	0	15	4	1	3	8	6	5
10	13	9	2	11	0	1	0	10	19	7	11	14	18	3

a)   Put the data into five equal classes.
b)   Lynn has already grouped the data she recorded in March. Find the modal class for each set of data. What can you say about the number of birds in March and June?

Lynn's data

Number of birds	0–3	4–7	8–11	12–15	16–19
Frequency	5	8	13	3	2

a)

Number of birds	0–3	4–7	8–11	12–15	16–19
Frequency	10	5	9	4	2

b)   The modal class for March was 8–11 and the modal class for June was 0–3. It seems likely that there were more birds in March than in June, because the modal class was higher.

Exercise 11.5

① Rupa counts the number of cars passing her house every 15 minutes.

Number of cars	0–9	10–19	20–29	30–39	40–49	50–59	60–69	70–79	80–89	90–99	100–109	110–119
Frequency	2	3	1	5	4	1	6	2	7	4	3	4

She decides that the frequencies for each interval are too small. Draw the frequency table again with intervals of 0–19, 20–39, and so on up to 100–119.

> Remember to add the frequencies for the two intervals together.

② A teacher asks his class how many minutes they spent revising for a maths test.

26	48	53	17	9	63	30	24	19	62	78	23
87	26	31	58	70	11	39	64	81			

Copy the frequency table below, and record this information in your table.

Time (minutes)	0–14	15–29	30–44	45–59	60–74	75–89
Tally						
Frequency						

3 A teacher has grouped the marks gained in a maths test by class 8F:

Mark	50–59	60–69	70–79	80–89	90–99
Tally	|||	卌	卌 |||	||||	||
Frequency	3	5	8	4	2

The marks gained by class 8G in the same test are:

> 62   61   75   55   86   70   76   51   68   92   87
> 95   57   68   60   59   96   81   77   61   95

a) Group the marks using a similar table.
b) What is the modal group for each of 8F and 8G?

> Look at the Example.

4 The numbers of spectators at different football matches were:

> 7263   8819   9586   3502   5265   4396   2348
> 3629   7201   5712   9063   8192   11 980   10 125

Copy and complete the table, filling in the two missing intervals.

Number of spectators	2000–3999	4000–5999			10 000–11 999
Tally					
Number of matches					

5 A P.E. teacher records the number of lengths swum by pupils in a form group.

> 2   6   18   9   14   10   6   15   7   24   11   18
> 12   8   19   22   20   8   11   23   17   25   16   13

a) Group this data using class intervals of 0–4, 5–9 and so on.
b) Group the data another way using different class intervals.

6 Jane records how long her family use the phone for.

Time on phone (seconds)	0–19	20–39	40–59
Number of phone calls			

Her mother makes a phone call lasting for 168 seconds – where will this be recorded?

7 What is wrong with these class intervals?

a)

Amount spent on a pair of trainers (£)	0–20	20–40	40–60	60–80	80+

b)

Amount spent on a pair of trainers (£)	0–19	20–39	40–59	60–79	80+

c) Suggest a correct way the class intervals could be written.

Investigation

8 This question requires an encyclopaedia or the Internet.

Collect your own data to compare two different regions. For example, investigate the longest rivers, tallest buildings, populations of cities or sports statistics.

Calculating statistics

⊕ Calculate statistics for small groups of discrete data

Key words
mean
median
mode
range
statistics

> The three most widely used averages are the **mean**, **median** and **mode**.
> You may need to choose an average that is most useful for representing the set of data.

Example 1   The marks given to a group of pupils in an examination are:

47   43   50   11   41   46   20   38   42   47   46   39   40   43   47

Find the mean, median, mode and range. Which average is most suitable for this data? Explain your answer.

Mean = 600 ÷ 15 = 40, median = 43, mode = 47,

range = 47 − 11 = 36

The median is the most suitable.

The mode has a frequency of just 2 and is the highest mark, so is

not typical; the mean is quite low compared to all of the marks.

Example 2   A shop sells tennis shoes in sizes 5, 6, 7, 8 and 9. The shopkeeper records sales of each size to find out which sell most quickly. He records the sales for one day in a frequency table:

Which average(s) should he use in deciding which size to re-order?

Size	Pairs sold
5	4
6	7
7	8
8	12
9	8

The mode (size 8) is the average that is most useful, as this

tells him which size will sell out first. The median (size 7) has

sold only just over half the number of pairs as the mode, and

the mean is 7.3, which does not exist as a shoe size.

Exercise 11.6

① A teacher takes a group of pupils on a school trip. The ages of the whole group are:

12   12   12   13   13   14   14   15   57

a) Calculate the median, mode and mean age of all of the group.

b) Calculate the median, mode and mean age not including the teacher's age.

c) Which of the averages stayed the same? Which average best represents the data?

2 Jane counts the number of raisins in 15 boxes.

  50   49   47   50   50   48   47   46   45   49   51   47   50   48   49

a) Calculate the three averages for the number of raisins in a box.

b) The average contents of a box is 50 raisins. Which of the three averages would this be?

3 The number of children in each year group in a small primary school are as follows:

Year	1	2	3	4	5	6
Number of children	28	31	30	25	31	29

The school would like to say that 'on average' class sizes were below 30.
Which of the three averages does **not** show this?

4 Each day Freya measures the temperature outside before going to work.

a) Calculate the three averages.

b) On Saturday the temperature is 0°C. Calculate the three averages for the six days.

c) Which of the three averages was most affected by the low temperature on Saturday?

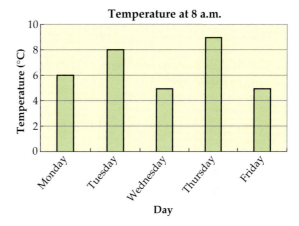

5 The ages of five children are 2 years, 2 years, 5 years, 6 years and 10 years.

a) Find the mean, median, mode and range of their ages now.

b) Find the mean, median, mode and range of their ages in one year's time.

c) Which of the four answers in part **b)** have remained the same? How have the others changed?

d) Use your answers from part **c)** to write down the mean, median, mode and range of their ages:

   i) in two years' time   ii) one year ago.

6 Think about information about a person that can only have a mode as an average.
Draw a table and write down five more types of data in each column.

Can only have a mode	Can have all three averages
hair colour	height

> Data that is not about number or measurement can only have a mode as an average.

Investigation

7 This question requires three coins.
Throw the three coins. Record the number of *heads* each time. Repeat the experiment 50 times.

Number of *heads*	0	1	2	3
Frequency				

Calculate the mode, median and mean number of *heads* each time.

> When finding the mean: remember to multiply the number of *heads* by frequency to find the total number of *heads*, then divide by 50.

Squares and square roots

◈ Recognise square numbers up to 100, and their square roots

◈ Use a calculator to find squares and square roots

Square numbers are found by multiplying a number by itself.
The square of 6 is 6×6 (or 6^2) = 36.
36 is a square number.

The **inverse** of squaring a number is finding its **square root**.
The square root of 36 (or $\sqrt{36}$) is 6.

The first 20 square numbers and their square roots are:

1^2	1	11^2	121
2^2	4	12^2	144
3^2	9	13^2	169
4^2	16	14^2	196
5^2	25	15^2	225
6^2	36	16^2	256
7^2	49	17^2	289
8^2	64	18^2	324
9^2	81	19^2	361
10^2	100	20^2	400

$7.3^2 \longrightarrow$ $\sqrt{240}$

To find 7.3^2, first use the table to find an estimate. The table shows that it is between 49 and 64.

Key in

[7] [.] [3] [x^2]

or

[7] [.] [3] [×] [7] [.] [3] [=]

on the calculator to give $7.3^2 = 53.29$

To find $\sqrt{240}$, the table shows that it is between 15 and 16.

Key in [2] [4] [0] [$\sqrt{}$] on the calculator to give:

$\sqrt{240} = 15.491\,933$

$\phantom{\sqrt{240}} = 15.5$ (rounded to 1 decimal place)

Example 1   What is 15.3^2?

Estimate: 15.3^2 = about 230

$15.3^2 = 234.09$

Check: 234.09 is close to estimate ✓

> It is between 15^2 and 16^2 but closer to 15^2.

> Use a calculator x^2 key or 15.3×15.3.

Example 2   The area of a square is 179.56 m². What is the length of one side?

Estimate: $\sqrt{179.56}$ is between 13 and 14

$\sqrt{179.56} = 13.4$

Check 13.4 is close to the estimate ✓

> Area of a **square** is found by $L \times L$ or L^2.
> To find one side we do the inverse (the square root).
> Area = 179.56 m²   Side = $\sqrt{179.56}$

> Use the $\sqrt{}$ key on the calculator.

Exercise 12.1

1 Using your table of squares, find the value of the following:

    a) 5^2 **b)** 18^2 **c)** 8^2 **d)** 14^2

    e) 3^2 **f)** 16^2 **g)** 17^2 **h)** 12^2

2 Use your table to find the following square roots.

a) $\sqrt{49}$ b) $\sqrt{400}$ c) $\sqrt{81}$ d) $\sqrt{121}$

e) $\sqrt{324}$ f) $\sqrt{16}$ g) $\sqrt{225}$ h) $\sqrt{4}$

3 Between which two whole numbers are the following square roots? Use your table to help. The first one has been done for you.

a) $\sqrt{175}$ is between 13 and 14

b) $\sqrt{38}$ c) $\sqrt{370}$ d) $\sqrt{92}$ e) $\sqrt{345}$

4 Between which two whole numbers are the following squares? Use the table to help. The first one has been done for you.

a) 18.2^2 – comes between 324 and 361

b) 4.5^2 c) 17.7^2 d) 15.72^2 e) 7.369^2

 5 Use your calculator to find the following squares. Remember to make an estimate first.

a) 6.7^2 b) 12.2^2 c) 16.9^2 d) 4.8^2

e) 3.2^2 f) 18.7^2 g) 14.1^2 h) 8.5^2

6 Copy and complete the following table. Give any decimal correct to 1 decimal place.

Number	A. Estimate $\sqrt{}$	B. Square Root	Difference C = B − A
$\sqrt{11}$			
$\sqrt{330}$			
$\sqrt{40}$			
$\sqrt{259}$			
$\sqrt{75}$			
$\sqrt{150}$			

7 Complete these calculations. Find each square number first and then do the calculation.

a) $2^2 + 5^2$ b) $7^2 - 3^2$ $2^2 + 5^2$ is not the same as $(2 + 5)^2$.

c) $16^2 - 14^2$ d) $13^2 - 9^2$

8 Work out the value of the following. Find the value of each square root first and then do the calculation.

a) $\sqrt{9} + \sqrt{16}$ b) $\sqrt{100} - \sqrt{25}$

c) $\sqrt{256} - \sqrt{144}$ d) $\sqrt{400} - \sqrt{64}$

 9 Ali has broken the square root key on his calculator.
He wants to find the $\sqrt{22}$. He tries to find it by this method:

He tries 4^2 to get 16.   Too small   He wants 22 so we need to try a larger number – say 5.

He tries 5^2 to get 25.   Too big   We now know the answer is between 4 and 5, so try 4.5.

He tries 4.5^2 to get 20.25   Too small

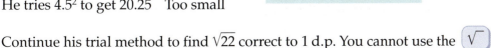

Continue his trial method to find $\sqrt{22}$ correct to 1 d.p. You cannot use the key.

Brackets

◈ Understand the meaning of a bracket

◈ Use the bracket keys on a calculator

◈ Know the order of mixed operations in a calculation

Some calculations have **brackets** . For example:

$(3 + 5) \times 2$

$58 - (4 + 3) \times 6$

A bracket tells you that the contents must be worked out first before any other **operations** .

5^2 and $(6 - 3)^2$ are examples of **powers** or **indices** .

The squared sign is sometimes called a power of 2.

The order for mixed operations in a calculation is:

Brackets

↓

Indices (or powers)

↓

Division and Multiplication

↓

Addition and Subtraction

So $(3 + 5) \times 2 = 8 \times 2 = 16$.

$58 - (4 + 3) \times 6 = 58 - 7 \times 6 = 58 - 42 = 16$.

Example 1   Without a calculator, find the value of:

a) $5 \times (3 + 2)$ **b)** $4^2 + 5 \times 2$

a) $5 \times (3 + 2)$

$= 5 \times 5$

$= 25$

b) $4^2 + 5 \times 2$

$= 16 + 5 \times 2$

$= 16 + 10$

$= 26$

Do the calculation in the brackets first.

No brackets, so we do indices first.

Multiplication comes before addition.

Example 2   Using the bracket keys on a calculator, work out $(3 + 9) \times (15 - 12)$

$(3 + 9) \times (15 - 12) = 36$

Key in:

(3 + 9) × (1 5 − 1 2) =

---

1 Find the value of the following:

a) $(3 + 2) \times 4$

b) $3 + 2 \times 4$

c) $(17 - 5) \times 3$

d) $(11 + 7) - (5 \times 3)$

e) $(5 + 4) \times (2 + 4)$

f) $9^2 + 4 \times 2$

2 Using the bracket key on a calculator, find the value of the following:

a) $6 \times (7 + 2)$

b) $(4.5 - 3.2) \times 8$

c) $(16 + 3) \times 4.3$

d) $19.6 - 2 \times 5$

e) $(132 - 79) \times 1.5$

f) $(21 + 4 + 7) \div 8$

g) $5 \times 3 \times 4 \div 12$

h) $81 \div (2^2 + 5)$

3 Find the answers to the calculations below. The answers are in the cloud, but an extra one has been included by mistake. Find the extra answer.

> 11, 28, 27, 24,
> 23, 22, 6, 4,
> 21, 18

$5 \times 4 + 3$ $4 \times 5 + 7$ $14 - 5 \times 2$

$4 \times 3 - 1$ $3 \times 7 + 3$ $12 - 3 \times 2$

$11 + 7 \times 1$ $7 + 5 \times 3$ $6 \times 5 - 2$

4 Copy the calculations below and put brackets in them to make them correct.

a) $4 + 5 \times 2 = 18$

b) $6 - 3 \times 2 = 0$

c) $4 \times 5 + 2 - 8 = 20$

d) $6 - 3 \times 9 - 4 = 15$

5 Copy and complete this cross number puzzle using the bracket key on a calculator.

> Remember to work from left to right.

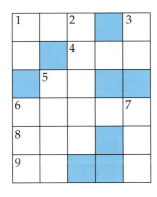

Down

1. $15 + 3 \times 2 - 2$

2. $35.3 \times 6 \div 2 \times 8$

3. $(11 - 3) \times 5$

5. $(5 + 7) \times (8 + 4) \times 10$

6. $32 \times (4^2 - 10)$

7. $5 \times (21 - 11) \times 5$

Across

1. $32 \times (5 + 3) \div 2$

4. $(7 + 3)^2 \times 4$

5. $544 \div (4 \times 8)$

6. 3.2×4.6

8. $(23.05 + 16.2) \times 48 \div 2$

9. $\sqrt{400}$

6 True or false?

a) $12 - 8 + 5 = 12 + 5 - 8$

b) $5 \times 8 - 3 = 8 - 3 \times 5$

c) $(4 + 3) \times 2 = 4 + (3 \times 2)$

d) $35 \div 5 \div 2 = 35 \div 2 \div 5$

e) $4 \times 9 \times 3 = 4 \times 3 \times 9$

⊕ Divide a 3-digit number by a 2-digit number using a standard written method

⊕ Estimate the result of a division by rounding

Dividing by a number is the same as a repeated subtraction of that number. The number you are dividing by is called a **divisor** .

To divide a number:

- make an **estimate** of the answer
- subtract as many 100's of the divisor as you can from the number
- subtract as many 10's of the divisor as you can from the result
- subtract as many units of the divisor as you can from the result
- write the **remainder** as a fraction in its lowest terms
- check that your answer is sensible by comparing it with the estimate.

Example Calculate $673 \div 42$

$10 \times 42 = 420, 20 \times 42 = 840$

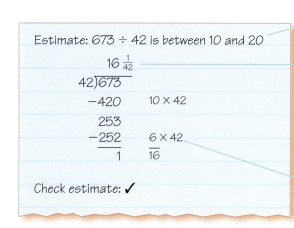

Estimate: $673 \div 42$ is between 10 and 20

$$16\frac{1}{42}$$
$$42\overline{)673}$$
$$-420 \quad 10 \times 42$$
$$253$$
$$-252 \quad 6 \times 42$$
$$\overline{1} \quad \overline{16}$$

Check estimate: ✓

1 remainder is written as a fraction $\frac{1}{42}$ i.e. 1 remainder when dividing by 42.

$10 \times 42 = 420$
$5 \times 42 = 210$
$210 + 42 = 252 = 6 \times 42$

Exercise 12.3

1 Complete these divisions:

a) $432 \div 4$ b) $582 \div 6$

c) $2635 \div 5$ d) $1603 \div 9$

Check your answer by multiplying your answer by the divisor. The result should match the starting number.

2 Complete these calculations.

a) 10×53 b) 20×32

c) 30×21 d) 30×25

e) 20×43

3 Complete these divisions. Remember to estimate your answer first.

    a) 294 ÷ 21   **b)** 678 ÷ 25

    c) 736 ÷ 32   **d)** 490 ÷ 18

    e) 787 ÷ 23   **f)** 585 ÷ 36

4 The cost of a 14-seater bus to the Ice Rink is £189.
How much is this for each person?

5 There are 888 bottles of beer to be packed into boxes
of 24 bottles.
How many boxes will there be?

> A fraction of a box will need to be counted as a whole box.

6 Jan has worked out that she delivered 924 newspapers to Valley Drive over 33 days.
How many is this per day?

7 Use a set of 1–9 digit cards.
Shuffle them and deal a 3-digit number and a 2-digit number.
Divide the larger number by the smaller one.
The whole number part of the answer is your score.
Shuffle and repeat for five rounds. Can you score a total of more than 50?

8 Sam earns £442 per year cleaning cars.
How much is this per week?

9 620 people live in an area of 16 square miles.
What is the mean number per square mile?

> We find the mean by dividing the number of people by the number of square miles.

12.4 Decimal division

Key words
divisor
nearest tenth
nearest hundredth

⊕ **Divide a decimal number by a 1-digit number using a standard written method**

⊕ **Estimate the result of a division by rounding**

Dividing by a number is the same as a repeated subtraction of that number. The number you are dividing by is called a **divisor** .

To divide a number:

- make an estimate of the answer
- subtract as many 100's of the divisor as you can from the number
- subtract as many 10's of the divisor as you can from the result
- subtract as many units of the divisor as you can from the result
- subtract as many tenths (0.1's) of the divisor as you can from the result
- subtract as many hundredths (0.01's) of the divisor as you can from the result
- check that your answer is sensible by comparing it with the estimate
- round the remainder to the **nearest** whole number, **tenth** or **hundredth** .

Example   Divide 457.6 by 4

```
Estimate: 457.6 ÷ 4 is about 120

          114.4
       4)457.6
        −400        100  × 4
          57.6
         −40         10  × 4
          17.6
         −16          4  × 4
           1.6
          −1.6       0.4 × 4
            0        114.4
Check estimate: ✓
```

480 ÷ 4 = 120

Exercise 12.4

1 Complete the following divisions.

a) 48.6 ÷ 3

b) 93.2 ÷ 4

c) 107.5 ÷ 5

d) 193.2 ÷ 6

e) 399.6 ÷ 3

2 Complete these divisions, writing your answer to 1 decimal place.

   a) $547.2 \div 4$

   b) $674.4 \div 3$

   c) $795.6 \div 6$

   d) $782.6 \div 5$

3 Choose any three parts from **Q1** or **Q2**.

For each one, multiply your answer by the divisor in the question part.

Check that the result is the same as the starting number in the question part.

4 125.7 cm of ribbon is to be cut into three pieces. How long will each piece be?

Give your answer rounded to the nearest whole number.

5 423.0 kg of sand is to be split into six large bags. How much will be in each bag?

Give your answer rounded to the nearest whole number.

6 Use a set of 2–9 digit cards.

Shuffle them and deal out four cards to make a number with 1 decimal place, for example 236.5. Now deal out your divisor card.

Divide your number, writing your answer to 1 d.p. where necessary.

Your score is the whole number part of your division. Can you score 500?

7 Peter ran five laps of an athletics track in 732.5 seconds.

   a) On average, how much is this for each lap?

   b) Now write your answer in minutes and seconds.

8 A tank holds 220.5 litres of water.

   a) How many 7-litre buckets can I fill?

   b) How much is left over?

Mental methods 1

⊕ **Use factors to make calculations simpler**

Finding a **factor pair** of a number is called **factorisation**.
For example, 35 can be factorised into 5×7.

Factorising a number in a **calculation** can sometimes make it easier.
Look for a factor of 10, for example:

$43 \times 20 \quad = 43 \times 2 \times 10$ Factorise 20 into 2×10.
$\qquad\qquad = 86 \times 10$
$\qquad\qquad = 860$

$3.2 \times 30 \quad = 3.2 \times 3 \times 10$ Factorise 30 into 3×10.
$\qquad\qquad = 9.6 \times 10$
$\qquad\qquad = 96$

If there is no factor of 10, look for a factor of 5, for example

$104 \times 15 \quad = 104 \times 3 \times 5$ Factorise 15 into 3×5.
$\qquad\qquad = 312 \times 5$
$\qquad\qquad = 312 \times 10 \div 2$ $\times 5$ is $\times 10$, then halve.
$\qquad\qquad = 3120 \div 2$
$\qquad\qquad = 1560$

Useful hints
To multiply by 5, multiply by 10, then halve.
To divide by 5, divide by 10, then double.

Example 1   Use factorisation to complete these multiplications.

    a) 23×30 **b)** 1.6×40 **c)** 224×15

a) $23 \times 30 = 23 \times 3 \times 10$
$\qquad\qquad\quad = 69 \times 10$
$\qquad\qquad\quad = 690$

Factorise 30 into 3×10.

b) $1.6 \times 40 \quad = 1.6 \times 10 \times 4$
$\qquad\qquad\qquad = 16 \times 4$
$\qquad\qquad\qquad = 64$

Factorise 40 into 10×4.

c) $224 \times 15 = 224 \times 3 \times 5$
$\qquad\qquad\quad = 672 \times 5$
$\qquad\qquad\quad = 672 \times 10 \div 2$
$\qquad\qquad\quad = 6720 \div 2$
$\qquad\qquad\quad = 3360$

Factorise 15 into 3×5.

$\times 10$ then halve is easier than $\times 5$.

Example 2   Use mental methods to complete these divisions.

    a) $480 \div 5$ **b)** $126 \div 14$

a) $480 \div 5 = 480 \div 10 \times 2$
$\qquad\qquad\quad = 48 \times 2$
$\qquad\qquad\quad = 96$

$\div 10$ then double is easier than $\div 5$.

b) $126 \div 14 = 126 \div 2 \div 7$
$\qquad\qquad\quad = 63 \div 7$
$\qquad\qquad\quad = 9$

Factorise 14 into 2×7.

Exercise 12.5

1 Use factorisation to complete these multiplications.

    a) 9×30 **b)** 6×80

    c) 45×20 **d)** 32×30

    e) 23×50 **f)** 42×40

Now check your answers with a calculator.

2 Use factorisation to complete these multiplications.

    a) 2.4×20 **b)** 1.3×30

    c) 3.2×40 **d)** 4.5×20

    e) 8.3×30 **f)** 9.2×40

Now check your answers using a calculator.

3 Copy and complete these number sentences.

    a) $28 \times 60 = 28 \times \square \times 10 = 168 \times \square = 1680$

    b) $5.8 \times 40 = 5.8 \times 4 \times \square = 23.2 \times \square = \square$

    c) $360 \div 30 = 360 \div 10 \div \square = 36 \div \square = 12$

    d) $640 \div 5 = 640 \div 10 \times \square = 64 \,\square\, 2 = 128$

    e) $35 \times 6 = \square \times 2 \times 3 = 70 \times \square = \square$

    f) $280 \div 35 = 280 \div \square \div 7 = 56 \div \square = 8$

4 Use mental methods to complete these divisions: | Look back at Example 2. |

    a) $130 \div 5$ **b)** $440 \div 5$ **c)** $42 \div 5$ **d)** $34 \div 5$ **e)** $244 \div 5$

Now check your answers using a calculator.

5 Use factorisation to complete these multiplications.

    a) 13×15 **b)** 25×14 **c)** 45×6 **d)** 13×9 **e)** 32×15

6 Use factorisation to complete these divisions.

    a) $480 \div 20$ **b)** $630 \div 30$ **c)** $600 \div 40$ **d)** $180 \div 60$ **e)** $210 \div 35$

7 There are 13 buses, each holding 40 people.
How many people is that altogether?

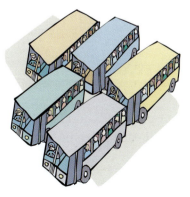

8 I have 300 tins of peas. How many boxes of 20 tins can I make?

9 John ran for 31 minutes. How many seconds is this?

Mental methods 2

◈ Use partitioning to make calculations simpler

Splitting a number into parts is called **partitioning** .

For example,   21 can be partitioned into $20 + 1$

439 can be partitioned into $400 + 39$

Partitioning a number in a **calculation** can sometimes make it easier.

Choose a partitioning which includes a multiple of 10, for example:

4.3×21 $= 4.3 \times 20 + 4.3 \times 1$ Partition 21 into $20 + 1$.

$= 86 + 4.3$

$= 90.3$

3.2×29 $= 3.2 \times 30 - 3.2 \times 1$ Partition 29 into $30 - 1$.

$= 96 - 3.2$

$= 92.8$

$295 \div 14$ $= 280 \div 14 + 15 \div 14$ Partition 295 into $280 + 15$.

$= 20 + 1\frac{1}{14}$

$= 21\frac{1}{14}$

Useful hints

To multiply by 50, multiply by 100, then halve.

Example 1   Using partitioning calculate:

a)   the cost of 14 jumpers at £21 each   b)   4.3 g of gold at £31 per gram.

a) 14×21 $= 14 \times 20 + 14 \times 1$ Split 21 into $20 + 1$.

$= 14 \times 2 \times 10 + 14 \times 1$

$= 280 + 14$

$= 294$

The cost is £294

b) 4.3×31 $= 4 \times 31 + 0.3 \times 31$ Split 4.3 into 4 and 0.3.

$= 4 \times 31 + 3 \times 31 \div 10$

$= 124 + 9.3$

$= 133.3$

The cost is £133.30

Example 2   A piece of wood is 169 cm long. If it is cut into 14 smaller pieces, how long will each piece be?

$169 \div 14 = (140 + 29) \div 14$ Split 169 into 140 and 29, as 140 is easy to divide by 14.

$= (140 \div 14) + (29 \div 14)$

$= 10 + 2\frac{1}{14}$

$= 12\frac{1}{14}$

Exercise 12.6

1 Copy and complete these number sentences to how you would partition these multiplications. The first one has been done for you.

    a) $6.3 \times 21 = 6.3 \times 20 + 6.3 \times 1$

    b) $2.3 \times 32 = 2.3 \times 30 +$

    c) $14 \times 51 = 14 \times 50 +$

    d) $22 \times 34 = \qquad + 22 \times 4$

    e) $4.2 \times 23 = \qquad + 4.2 \times 3$

2 Now complete the calculations for the multiplications in **Q1**.

3 Use partitioning to complete these multiplications:

    a) 23×31 **b)** 32×23 **c)** 24×42 **d)** 45×21

4 Copy and complete these number sentences. The first one has been done for you.

    a) $294 \div 14 = 280 + 14 \div 14 = 280 \div 14 + 14 \div 14$

    b) $252 \div 21 = 210 + 42 \div 21 = 210 \div 21 +$

    c) $208 \div 16 = 160 + 48 \div 16 = 160 \div 16 +$

    d) $276 \div 23 = 230 + 46 \div 23 = \qquad + 46 \div 23$

    e) $416 \div 32 = 320 + 96 \div 32 = \qquad +$

    f) $473 \div 43 = 430 + 43 \div 43 = \qquad +$

5 Now complete the divisions in **Q4**.

6 Use partitioning to divide each number by 21.

    a) 483 ———————————————— 483 partitions into $420 + 63$.

    b) 672

    c) 903

7 A builder uses 176 metres of curtain rail in 16 houses. How much is that per house?

8 Each box of toffees contains 45 sweets. If I have 22 boxes, how many sweets is that altogether?

9 A wren has a mass of 2.1 g. What is the mass of 43 wrens?

13.1 The general term

Key words
term
general term
*n*th term

⊕ Generate a sequence given the general term

You can generate a sequence of numbers if you are given the first **term** and the term-to-term rule.

For example, if the first term is 4 and the term-to-term rule is 'add 3' the sequence is: 4, 7, 10, 13, 16, …

You can also generate a sequence from the **general term**. This is sometimes called the ***n*th term**. It is a formula for calculating the value of any term in the sequence.

Example 1 Find the first five terms of a sequence whose general term is $2n + 1$.

We can generate the sequence by substituting values for *n* in the general term.

Substitute: $n = 1$ to find the 1st term

$n = 2$ to find the 2nd term

$n = 3$ to find the 3rd term

$2 \times \mathbf{1} + 1 = 3$: Substitute $n = 1$ into $2n + 1$

and so on.

Term number (n)	1	2	3	4	5
Sequence	$2 \times 1 + 1 = 3$	$2 \times 2 + 1 = 5$	$2 \times 3 + 1 = 7$	$2 \times 4 + 1 = 9$	$2 \times 5 + 1 = 11$

Example 2 The nth term of a sequence is $3n + 5$.
  a) Find the 23rd term.   **b)** Find the 16th term.

a) $3 \times 23 + 5 = 74$ — To find the 23rd term, substitute $n = 23$ into the expression $3n + 5$.

b) $3 \times 16 + 5 = 53$ — To find the 16th term, substitute $n = 16$.

Exercise 13.1

1 Find the first five terms of the sequences whose general term is:
  a) $n + 5$ **b)** $4n$ **c)** $5n - 2$ **d)** $4n + 8$ **e)** n

2 Find the 25th terms of the each of the sequences in **Q1**.

3 Find the 17th term of a sequence whose general term is $4n + 5$.

4 The *n*th term of a sequence is $3n - 5$.
  a) Find the 9th term.   **b)** Find the 25th term.
  c) Find the 50th term.   **d)** Find the 100th term.

5 The number of bricks needed to build houses depends on the number of houses joined in a row.

A builder uses the general term: $2000n + 1000$ to calculate the number of bricks needed for any number of houses.

How many bricks will he need for:

a) two houses

b) five houses

c) ten houses?

Substitute the number of houses into the general term.

6 A taxi driver calculates the cost of each journey in pounds using this general term: $0.5n + 2$ where n is of the distance in miles.

What is the cost for a journey of:

a) 10 miles

b) 15 miles

c) 25 miles?

7 The general term of a sequence is $3n + 2$.
One of the terms in the sequence is 17.

a) Find the term number (n) when the term is 17 by solving the equation $3n + 2 = 17$

b) Find the term number (n) when the term is:

   i) 5

   ii) 77

   iii) 302

Use inverse operations or the balancing method.

8 Sarah makes curtains for windows.
She uses the formula $2n + 0.5$, where n is the width of the window, to work out how much material she needs in metres.
Sarah uses 6.5 m of material for a pair of curtains.
What was the width of the window?

Solve the equation
$2n + 0.5 = 6.5$

Investigation

9 a) Write down the first five terms of the sequence whose general term is $2n$.

b) What are the differences between consecutive terms?

c) Write down the first five terms of the sequence whose general term is $2n + 5$.

d) What are the differences between consecutive terms?

e) What do you notice about the differences between consecutive terms and the general term?

Exploring sequence patterns

⊕ Find the general term of a sequence of patterns

Look at this **sequence** :

⊗
⊗ ⊗

⊗
⊗ ⊗
⊗ ⊗

⊗
⊗ ⊗
⊗ ⊗
⊗ ⊗

⊗
⊗ ⊗
⊗ ⊗
⊗ ⊗
⊗ ⊗

We can see how this sequence grows by looking at the number of ⊗ in each pattern.
The number of ⊗ increases by 2 each time.

⊗
⊗ ⊗

⊗
⊗ ⊗
⊗ ⊗

⊗
⊗ ⊗
⊗ ⊗
⊗ ⊗

The 1ˢᵗ pattern
has **1** lot of 2⊗
plus 1 more ⊗

The 2ⁿᵈ pattern
has **2** lots of 2⊗
plus 1 more ⊗

The 3ʳᵈ pattern
has **3** lots of 2⊗
plus 1 more ⊗

Following this pattern we can see that:
The **10**ᵗʰ pattern will have **10** lots of 2⊗ plus 1 more ⊗
The **100**ᵗʰ pattern will have **100** lots of 2⊗ plus 1 more ⊗.
The nth pattern will have n lots of 2⊗ plus 1 more ⊗.
So the nth term is $n \times 2 + 1 = 2n + 1$.
This is the **general term** of the sequence.

Example

a) Draw the next two patterns in the sequence below:

◇ ◇
◇ ◇ ◇

◇ ◇
◇ ◇ ◇
◇ ◇ ◇

◇ ◇
◇ ◇ ◇
◇ ◇ ◇
◇ ◇ ◇

◇ ◇
◇ ◇ ◇
◇ ◇ ◇
◇ ◇ ◇
◇ ◇ ◇

b) Describe in words how the sequence is growing.
c) Describe the 50ᵗʰ term.   d) Find the nᵗʰ term.

a)

◇ ◇
◇ ◇ ◇
◇ ◇ ◇
◇ ◇ ◇
◇ ◇ ◇
◇ ◇ ◇

◇ ◇
◇ ◇ ◇
◇ ◇ ◇
◇ ◇ ◇
◇ ◇ ◇
◇ ◇ ◇
◇ ◇ ◇

b) The number of ◇ increases by 3 each time.
c) The 1ˢᵗ term has **1** lot of 3 ◇ plus 2 more ◇.
     The 2ⁿᵈ term has **2** lots of 3 ◇ plus 2 more ◇.
     The 3ʳᵈ term has **3** lots of 3 ◇ plus 2 more ◇.
     The 50ᵗʰ term has **50** lots of 3 ◇ plus 2 more ◇.
d) The nth term has n lots of 3 ◇ plus 2 more ◇.
     So the nth term is $n \times 3 + 2 = 3n + 2$.

Look at the sequence. We can
see that each time three more ◇
are added.

Exercise 13.2

1 **a)** Draw the next two terms of this sequence:

b) Describe in words how the pattern is growing.
c) Describe the 7th term.
d) Describe the 10th term.
e) Describe the 20th term.
f) Find the general term.

> See part **d)** of the Example.

2 For each of the sequences below:
a) Describe how the sequence is growing
b) Describe the 10th and 50th terms.
c) Find the nth term.

i)

ii)

3 For each of the pattern sequences below, find the general term and use it to calculate the number of dots in the 10th term.

a)

b)

4 Draw dot patterns for each of the sequences below. The first one is started for you.

a)

Term number	1st	2nd	3rd	4th	5th	6th
Sequence	2	4	6			
Pattern	• •	• • • •				

b)

Term number	1st	2nd	3rd	4th	5th	6th
Sequence	3	6	9			

c)

Term number	1st	2nd	3rd	4th	5th	6th
Sequence	10	20	30			

5 Find the general term for each of the sequences in **Q4**.

13.3 Finding the general term

Find the general term of a sequence

Key words
general term
sequence
consecutive terms
nth term

To find the **general term** of a **sequence**, look at how the sequence grows.

In the sequence **4, 7, 10, 13, 16,** … the numbers
grow by 3 each time.
We can write the sequence in a table:

Term number	1	2	3	4	5
Sequence	4	7	10	13	16
	3 + 1	3 + 3 + 1	3 + 3 + 3 + 1	3 + 3 + 3 + 3 + 1	3 + 3 + 3 + 3 + 3 + 1

The **1st** term is **1** lot of 3 plus 1 more
The **2nd** term is **2** lots of 3 plus 1 more
The **3rd** term is **3** lots of 3 plus 1 more
and so on.

The difference between **consecutive terms** is 3, so we compare the sequence with the 3× table plus 1.

The **nth term** is n lots of 3 plus 1 more $= n \times 3 + 1 = 3n + 1$

Example   Find the general term for the number of dots in the sequence below:

Two more dots are being added each time. We can compare the sequence with the 2× table plus 3.

The **1st** term is **1** lot of 2 plus 3 more

The **2nd** term is **2** lots of 2 plus 3 more

The **3rd** term is **3** lots of 2 plus 3 more

The **nth** term is **n** lots of 2 plus 3 more $= n \times 2 + 3 = 2n + 3$

Exercise 13.3

1 Look at this sequence:

a) How many dots are being added each time?
b) Copy and complete:
   The 1st term is 1 lot of ☐ plus ☐ more
   The 2nd term is 2 lots of ☐ plus ☐ more
   The 3rd term is 3 lots of ☐ plus ☐ more
   The nth term is n lots of ☐ plus ☐ more $= n \times$ ☐ $+$ ☐ $=$ ☐$n +$ ☐

156   Maths Connect 2G

2 a) Copy and complete the table below:

Term number	1	2	3	4	5
Sequence	5 $\square$ + 1	9 $\square$ + $\square$ + 1	13	17	21

b) Use your table to find the nth term of the sequence.

See the Example.

3 Find the general term for each of these sequences by drawing tables as in **Q2**.

a) 7, 10, 13, 16, 19, …   **b)** 5, 8, 11, 14, 17, …

4 Find the general term for each of these sequences:

a)

Term number	1	2	3	4
Sequence	9	16	23	30

b)

Term number	1	2	3	4
Sequence	£10	£18	£26	£24

c) 8, 14, 20, 26, 32, … …
d) 5 cm, 7 cm, 9 cm, 11 cm, 13 cm, …

5 Find the general term for the number of squares in each of the following patterns:

a)   **b)**

Investigations

6 Write down the first five terms of the following sets of sequences.

a) $3n + 5$, $3n - 2$, $3n$ **b)** $4n - 4$, $4n$, $4n + 3$
c) $10n$, $10n + 100$, $10n - 5$

What do you notice about the difference between consecutive terms for the sequences in part **a)**? Does the same thing happen with the sequences in parts **b)** and **c)**?

Suggest how you could find the number the n is multiplied by in the general term, by looking at the differences between consecutive terms.

7 If the general term of a sequence is $2n + 1$ then the first few terms of the sequence are: 3, 5, 7, 9, …
We could also describe this sequence by saying: First term = 3   term to term rule = +2

a) Describe each of these sequences by giving the first term and term-to-term rule:

  i) $2n + 3$ **ii)** $3n - 2$ **iii)** $5n + 10$ **iv)** $4n - 3$

b) What do you notice about the relationship between the nth term and the term-to-term rule?

c) What do you notice about the relationship between the nth term and the first term?

You should check your answers to parts **b)** and **c)** using other sequences.

Spreadsheets

⊕ Use spreadsheets to generate sequences

Key words
column
row
cell
generate

A spreadsheet page looks like this.

- **Columns** go down the page and **rows** go across.
- Each individual space in a spreadsheet is called a **cell**. The area shaded red in this spreadsheet is one of the cells.
- The area shaded red is cell B5. You always write the column letter before the row number.
- You can enter information into a spreadsheet by moving to a particular cell and typing values in (see cell A8).

	A	B	C	D	E
1					
2				= 5/7	
3					
4					
5					=A8×2+50
6					
7					
8	29				
9					
10					

- If you want the spreadsheet to calculate a value for you enter '=' and then the calculation you want it to do (see cell D2).
- You can enter calculations in cells that refer to other cells. The formula in cell E5 tells the computer to multiply the amount in cell A8 by 2 and then to add 50. Spreadsheets can be used to **generate** sequences.

Example   Use a spreadsheet to generate the sequence: 15, 20, 25, 30, 35, ...

	A	B	C
1	15	1	=5×B1+10
2	=A1+5	2	=5×B2+10
3	=A2+5	3	=5×B3+10
4	=A3+5	4	=5×B4+10
5	=A4+5	5	=5×B5+10
6	=A5+5	6	=5×B6+10
7	=A6+5	7	=5×B7+10
8	=A7+5	8	=5×B8+10
9	=A8+5	9	=5×B9+10
10	=A9+5	10	=5×B10+10

Method 1: Describe the sequence by giving the first term 15 and a term-to-term rule + 5.
Enter the first term in the sequence (15) in cell A1. The second term in the sequence is the first term + 5 so enter '= A1 + 5' into cell A2 and so on.
Terms in the sequence are described in relation to the previous term.

Method 2: If you know the general term of the sequence, you can use this to generate the sequence. The general term of the sequence 15, 20, 25, 30, 35, ... is $5n + 10$.
Enter the term numbers into column B.
Each term in the sequence is: 5 × term number + 10
Enter '= 5 × B1 + 10' into cell C1 and so on.
Terms in the sequence are described in relation to the term numbers.

Exercise 13.4

1 The first few terms of a sequence are 51, 101, 151, 201, ...

a) Describe the sequence by giving the first term and the term-to-term rule.

b) Use a spreadsheet package to generate the sequence as far as the 10th term, using the information from part **a)**.

> See Method 1 in the Example.

c) The general term of the sequence is $50n + 1$. Use the general term to generate the sequence on a spreadsheet package as far as the 15th term.

> See Method 2 in the Example.

2 By finding the first term and the term-to-term rule, generate the first ten terms of these sequences on a spreadsheet:

a) 14, 19, 24, 29, 34, …   b) 8, 12, 16, 20, 24, …

c) 1, 1.5, 2, 2.5, 3, 3.5, …   d) 113, 116, 119, 122, …

3 Generate the first fifteen terms of each of the following sequences on a spreadsheet:

a) $6n + 3$ b) $100n + 5$

c) $0.5n + 2$ d) $5n - 2$

4 a) Write down the sequence of numbers generated when the following information is put into a spreadsheet:

b) What is the first term of the sequence?

c) What is the term-to-term rule?

d) Find the general term of the sequence.

e) The same sequence of numbers is generated in column C below using the general term of the sequence. Write down the formulae you would enter in cells C1 to C10.

	A	B
1	−1	
2	=A1+5	
3	=A2+5	
4	=A3+5	
5	=A4+5	
6	=A5+5	
7	=A6+5	
8	=A7+5	
9	=A8+5	
10	=A9+5	

	A	B	C
1	−1	1	
2	=A1+5	2	
3	=A2+5	3	
4	=A3+5	4	
5	=A4+5	5	
6	=A5+5	6	
7	=A6+5	7	
8	=A7+5	8	
9	=A8+5	9	
10	=A9+5	10	

5 The first three terms of some sequences are generated in a spreadsheet using the general terms in columns B, C and D. The spreadsheet is shown:

	A	B	C	D
1	1	=7×A1+5	=5×A1+5	=A1−3
2	2	=7×A2+5	=5×A2+5	=A2−3
3	3	=7×A3+5	=5×A3+5	=A3−3

Write down:

a) The general term of each sequence.

b) The first five terms of each sequence.

c) The first term and the term-to-term rule for each sequence.

Investigation

6 Use a spreadsheet package to investigate ways of generating these sequences:

a) Multiples of 12

b) The even numbers

c) The odd numbers

d) The square numbers.

13.5 Conversion graphs

⊕ Read values from a conversion graph
⊕ Plot a conversion graph

Key words
variable
plot
conversion graphs

We can show a relationship between two **variables** by **plotting** a graph.

We can use a **conversion graph** to help us convert from one unit to another.

Example   The following graph converts distances in inches into distances in cm.

     a)   A ruler is 10 inches long. What is this in cm?

     b)   A pencil is 20 cm long. What is this in inches?

     c)   Explain why the graph goes through the origin.

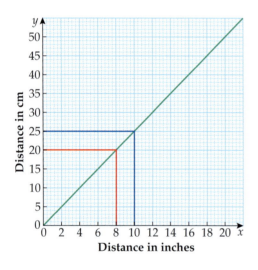

Draw a vertical line up from 10 inches on the x-axis to the graph and a horizontal line from the meeting point on the graph to the y-axis. Read off the length in cm.

Draw a horizontal line from 20 cm on the y-axis to the graph and a vertical line from the meeting point on the graph to the x-axis. Read off the length in inches.

a)   The ruler is 25 cm long.

b)   The pencil is 8 inches long.

c)   A distance of 0 cm is equal to a distance of 0 inches, so the graph goes through the origin.

The origin is the point (0, 0).

Exercise 13.5

1 **a)** Use the graph in the Example to convert the following lengths into inches:

      i) 5 cm   **ii)** 20 cm   **iii)** 50 cm   **iv)** 35 cm

    b) Use the graph in the Example to convert the following lengths into cm:

      i) 2 inches   **iii)** 10 inches   **iii)** 20 inches   **iv)** 1 inch

2 A waiter uses the following graph to work out how much he earns each day.

Read the values off the graph.

a) Copy and complete the table below to show how much the waiter earns each day:

Day	Number of hours worked	Amount earned
Monday	5 hours	......
Tuesday	10 hours	......
Wednesday	7 hours	......
Thursday	......	£20
Friday	......	£45
Saturday	......	£15
Sunday	0 hours	......

b) How much does the waiter earn this week?

c) How many hours did the waiter work for during the week?

d) Explain why the graph goes through the origin.

3 A bank displays the following information to help convert between pounds and euros:

Pounds (£)	Euros (€)
20	28
50	70
150	210

a) On graph paper draw a pair of axes with pounds (£) on the x-axis and euros (€) on the y-axis.

b) Plot the three points from the table and join them with a straight line.

The x-axis goes up to £150. The y-axis goes up to 210 euros.

c) What is £33 in euros?

d) How much is 25 euros in pounds?

e) Use your graph to decide whether the following statements are true or false.
   i) £100 is the same as 140 euros.
   ii) If I change £30 into euros, I get 50 euros.
   iii) If I change 300 euros back into pounds, I get £200.
   iv) £1 is worth less than €1.

4 Look at all the graphs you have used in this exercise.
   a) Do they all go through the origin?
   b) Why is this?

Investigation

5 We can measure temperature in two different ways: either in degrees Fahrenheit (°F) or degrees Celsius (°C).
   a) Find out the temperature in degrees Fahrenheit that is equivalent to 0°C.

   Use the Internet or look in a science book.

   b) If you drew a conversion graph for converting from degrees Celsius to degrees Fahrenheit would it go through the origin? Explain your answer.

Drawing graphs

⊕ Draw graphs to illustrate real life situations

Key words
graph

By drawing a **graph** we can display a lot of information in a small space.

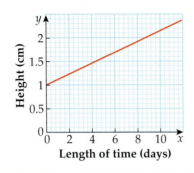

This graph shows the height in cm of a plant in a greenhouse over a period of 10 days. It tells you how tall it was after 1 day, 2 days, 3 days and so on. You can read these values off the graph.

Example   GoGas charges a basic monthly charge of £5, as well as £0.20 for each unit of gas used.

     a) Draw up a table and calculate the cost when 0, 10 and 50 units of gas are used.

     b) Draw a graph to show the cost of the monthly gas bill.

a)

Number of units used	0	10	50
Cost in pounds	$0.20 \times 0 + 5 = 5$	$0.20 \times 10 + 5 = 7$	$0.20 \times 50 + 5 = 15$

b)

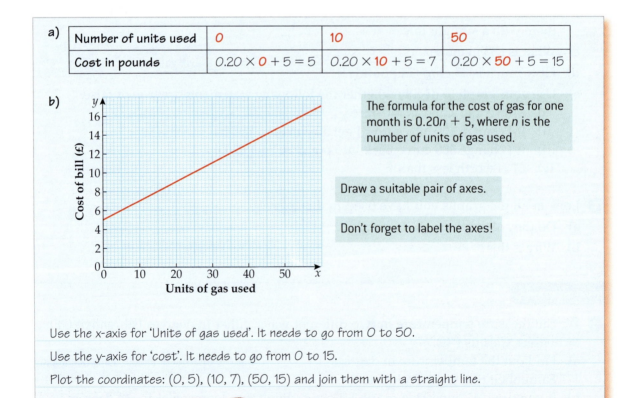

The formula for the cost of gas for one month is $0.20n + 5$, where n is the number of units of gas used.

Draw a suitable pair of axes.

Don't forget to label the axes!

Use the x-axis for 'Units of gas used'. It needs to go from 0 to 50.

Use the y-axis for 'cost'. It needs to go from 0 to 15.

Plot the coordinates: (0, 5), (10, 7), (50, 15) and join them with a straight line.

Exercise 13.6

1 Spark-elec supplies electricity.

It charges a standing charge of £10 per month and £0.10 for each unit used.

a) Copy and complete the table of values below:

Number of units used	0	10	20	30
Cost in £				

Write a formula for the cost in pounds each month.

b) Draw a pair of axes from 0 to 30 on the *x*-axis and 0 to 25 on the *y*-axis.

c) Label the *x*-axis 'Number of units used' and the *y*-axis 'Cost (£)'.

d) Plot the coordinates from your table and join the points with a straight line.

e) Use your graph to work out the cost of a bill when:

    i) 5 units   **ii)** 15 units   **iii)** 22 units are used

f) Use your graph to work out the number of units of electricity used if the bill costs:

    i) £22.50   **ii)** £11.50   **iii)** £16

2 A car travels at a constant speed of 30 miles per hour.

a) Copy and complete the table of values below:

Time (hours)	Distance travelled (miles)
1	…
2	…
…	90
…	120

b) Draw a graph to show this information. Draw the *x*-axis up to 5 hours.

c) Extend the line of your graph to show how far the car would travel in 5 hours.

d) Use your graph to work out how far the car would travel in:

    i) $2\frac{1}{2}$ hours   **ii)** $4\frac{1}{2}$ hours   **iii)** $1\frac{1}{4}$ hours

e) How long would it take for the car to travel:

    i) 45 miles   **ii)** 105 miles   **iii)** 0 miles?

3 A mobile phone company offers different tariffs.
The 'All Talk' tariff costs 50p per minute with no line rental.
Plot a graph to show the cost of the monthly bill if you
made between 0 and 60 minutes worth of calls in one month.

Use the method you have used for **Q1–2**.

4 Look at your graphs for **Q1–3**.

a) Which ones go through the origin?

b) Explain why you think this is.

c) Why doesn't the other graph go through the origin?

5 Find out about different tariffs for mobile phones.
Draw separate graphs to show the different costs
of the monthly bills for between 0 and 60 minutes
worth of calls.
Compare the graphs and discuss which you think has the best deal.

You could work in pairs or groups, each group drawing the graph for a different tariff.

Interpreting graphs

⊕ Discuss and describe the shape of real life graphs

Graphs show the relationship between two **variables** .

We can **interpret** the relationship between the two variables by looking at the slope of the graph and the **y-intercept** .

For this graph, as the *x*-value increases the *y*-value increases.

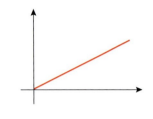

The *y*-intercept tells us the value of *y* when $x = 0$.

For this graph, as the *x*-value increases the *y*-value decreases.

Example   The following graph shows the number of birds in Umabungo over a period of ten years from 1990 to 2000.

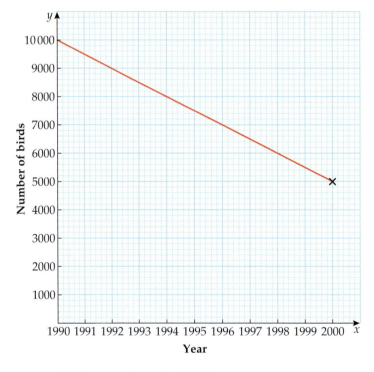

a) How many birds were there in 1990?
b) Over the ten years what happened to the number of birds?
c) How is this shown on the graph?

a) There were 10 000 birds. ———————————————— The *y*-intercept is 10 000.

b) The number of birds decreased from 10 000 to 5000.

c) On the graph, as the x-value increases the y-value decreases.

Exercise 13.7

1 The graphs show the annual rainfall in different countries.

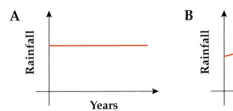 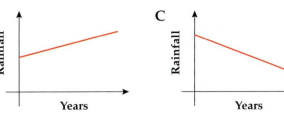

Match each graph with one of the descriptions below:
- **i)** The amount of rainfall has increased over the past few years.
- **ii)** The amount of rainfall has decreased over the past few years.
- **iii)** The amount of rainfall has remained the same over the past few years.

2 Electricity is measured in units.
The following graph shows the cost of an electricity bill according to the number of units used.

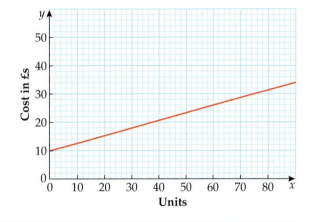

- **a)** What is the cost of the bill for a family that uses 20 units?
- **b)** If the electricity bill is £25, how many units were used?
- **c)** What does the slope of the graph tell us?
- **d)** What would the cost be for a family that used 0 units?
- **e)** What is the standing charge? The standing charge is the cost if you use no units of electricity.

3 A café manager drew the following graph to show the number of mugs of hot chocolate sold against the temperature outside.

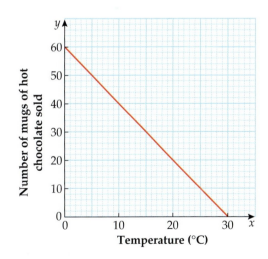

- **a)** What does the slope of the graph tell us?
- **b)** What is the y-intercept?
- **c)** What does the y-intercept tell us about the amount of hot chocolate sold when the temperature is 0°C?

4 Draw rough sketches of graphs to illustrate the following situations:
- **a)** The speed of a car as it accelerates from stationary, plotted against time.
- **b)** The mass of a burning candle against time.
- **c)** The number of pages in this textbook against time.
- **d)** Your height between the ages of 20 and 60.
- **e)** Your height between the ages of 0 and 20.

⊕ Draw and interpret graphs from real life

This graph shows the percentage of households in the UK with Internet access from 1998 to 2003.

We can see that the percentage of households with Internet access increased from 1998 to 2003.

The percentage increased by a large amount between 1998 and 2000.

Since 2000 the increase has continued, but more slowly.

The steeper the **gradient** of the graph, the faster the increase.

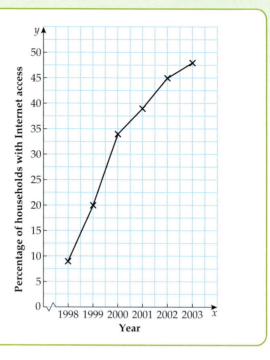

Example Mr Murphy goes for a run each morning. He sets off from home at 9.00 am.

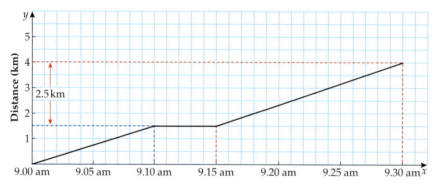

a) How far is Mr Murphy from home after 10 minutes?
b) What is Mr Murphy doing between 9.10 am and 9.15 am?
c) How far does Mr Murphy run in the last 15 minutes of his jog?
d) How far does Mr Murphy run in total?
e) How long does Mr Murphy run for?

See blue dotted line.

a) After 10 minutes Mr Murphy is 1.5 km from home

b) Mr Murphy is resting ——— The line is horizontal so he is not going anywhere.

c) 2.5 km ——— See the red dotted lines.

d) 4 km ——— The highest point the graph reaches is 4 km.

e) 30 minutes ——— Mr Murphy starts his run at 9.00 am and finishes at 9.30 am.

Exercise 13.8 ....

1 The graph opposite shows Mr Key's car journey from Petersfield to Oxford:

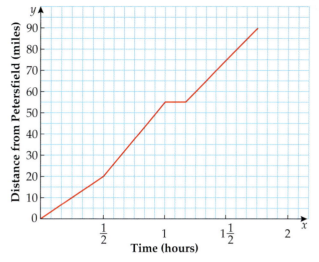

a) How far was Mr Keys from Petersfield after driving for 1 hour? How far was he from Oxford at this time?

b) How far was Mr Keys from Petersfield after driving for $1\frac{1}{2}$ hours?

c) What did Mr Keys do 1 hour into his journey?

d) How far is Oxford from Petersfield?

e) How long did it take Mr Keys to drive from Petersfield to Oxford?

2 The following graph shows two swimmers racing over a distance of 800 m.

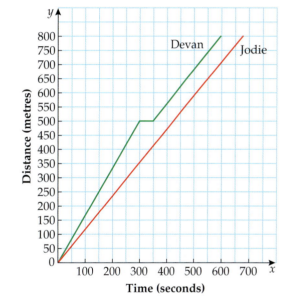

a) How far does Devan swim in 600 seconds?

b) How far does Jodie swim in 600 seconds?

c) What does Devan do during the race that Jodie doesn't?

d) Who wins the race?

3 Mr James is taking part in a sponsored walk.
He starts at 3.30 pm and walks 1 km in the first 10 minutes.
He then stops for a 2 minute rest.
Finally he walks 2 km in the next 18 minutes.

a) Draw a graph to show how far Mr James is from the start at different times.

> Show the time on the x-axis and distance in km on the y-axis.

b) How far does Mr James walk in total?

4 Emily swims 30 lengths of her local pool in 30 minutes.

a) The pool is 30 m long. How far does she swim in total?

b) She swims at a steady rate of 1 length per minute. Draw a graph to show her swim.

> Label the x-axis 'Time in minutes' and the y-axis 'Distance in metres.'

c) Deborah also swims 30 lengths in 30 minutes.
First she swims 10 lengths in 8 minutes and then takes a 2 minute break.
She then swims the remaining 20 lengths.
On the same pair of axes, draw a graph to illustrate Deborah's swim.

⊕ Understand the meaning of proportion

⊕ Solve problems involving proportion

> If   £1 is worth 2.5 A$ (A$ – Australian dollars)
> then   £2 is worth 5.0 A$
>   £4 is worth 10.0 A$, and so on.
>
> The pound is in **direct proportion** to the Australian dollar.
>
> When things are in direct proportion to each other, we can solve problems about them, using the method of finding the value of one of them.
>
> For example:
> The cost of tickets is in direct proportion to the number of tickets bought.
>
> If 4 tickets cost £8.40, then we can find the cost of 5 tickets, or any other number of tickets, by first finding the cost of 1 ticket.
>
> $\div 4 \left(\begin{array}{l}\text{4 tickets cost £8.40} \\ \text{1 ticket costs £2.10}\end{array}\right) \div 4$
>
>   5 tickets cost £8.40 + £2.10 = £10.50

Example 1   If 8 tickets to the cinema cost £24, what is the cost of 12 tickets?

> 8 tickets cost £24
>
> 4 tickets cost £24 ÷ 2 = £12
>
> 12 tickets cost £24 + £12 = £36

Example 2   The ingredients for 10 chocolate crispie cakes are:

      400 g   Rice Crispies
      250 g   Chocolate
      6 teaspoons   Golden Syrup

How much of each ingredient would I need to make 15 crispie cakes?

> *Rice Crispies*
> 10 cakes need 400 g
> 5 cakes need 400 ÷ 2 = 200 g
> 15 cakes need 400 + 200 g = 600 g
>
> *Chocolate*
> 10 cakes need 250 g
> 5 cakes need 250 ÷ 2 = 125 g
> 15 cakes need 250 + 125 g = 375 g
>
> *Golden Syrup*
> 10 cakes need 6 teaspoons
> 5 cakes need 6 ÷ 2 = 3 teaspoons
> 15 cakes need 6 + 3 = 9 teaspoons

Exercise 14.1

1 Copy and complete this direct proportion table to find the cost of chocolate eggs costing £3.10 each.

Number	Cost
1	£3.10
2	£6.20
3	
4	
5	
6	
7	
8	
9	
10	

2 If three pies cost £6.30 find the cost of:
    a) 6 pies   **b)** 9 pies.

3 500 g of mince is needed to make a Shepherds Pie for 4 people.
    a) How much mince would I need for 1 person?
    b) How much mince would I need for 5 people?

> Look at Example 1.

4 A recipe for a fruit cocktail is:
This is enough for 4 people.
How much would I need for 5 people?

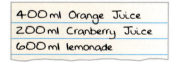

```
400 ml Orange Juice
200 ml Cranberry Juice
600 ml lemonade
```

5 One tin of purple paint is made up of 600 ml of blue paint and 400 ml of red paint. How much of each colour would I need to make 7 tins of paint?

6 Four ice-lollies cost £1.60.
    a) What is the cost of 2 ice-lollies?
    b) What is the cost of 8 ice-lollies?
    c) How much change would I get from £20, if I bought 8 ice-lollies?

7 A recipe for two small cakes is:
How much of each ingredient would I need for
    a) 4 cakes
    b) 6 cakes?

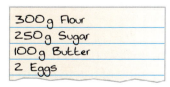

```
300 g Flour
250 g Sugar
100 g Butter
2 Eggs
```

8 £1 is worth approximately 1.5 euros.
    a) How many euros would I get if I changed £16 into euros?
    b) I have 15 euros. How much is this worth in pounds?

Ratio and proportion

- Understand the meaning of ratio
- Reduce a ratio to its simplest form
- Solve problems involving ratio

A **proportion** compares one part with the whole.

A **ratio** compares one part with another part.

The proportion of red cards is $\frac{7}{10}$.
The proportion of blue cards is $\frac{3}{10}$.
The ratio of red : blue is 7 : 3.
The ratio of blue : red is 3 : 7.

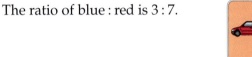

A ratio can compare more than two parts.
The ratio of flowers : cars : mushrooms : shells is 4 : 2 : 3 : 1.

The ratio of cars : non cars 2 : 8.
The ratio 2 : 8 simplifies to 1 : 4 by dividing both sides by 2.
This is like simplifying a fraction by **cancelling** .

Example 1

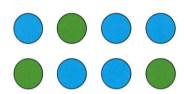

a) What **proportion** of the counters are blue?
b) What **proportion** of the counters are green?
c) What is the **ratio** of green counters to blue counters?
d) What is the **ratio** of blue counters to green counters?

a) $\frac{5}{8}$ b) $\frac{3}{8}$

> Proportion compares number of green counters to the whole set of counters.

c) 3 : 5   d) 5 : 3

> Ratio compares number of blue counters to number of green counters – remember the order matters.

Example 2

In a wood there were 18 ash trees, 12 oak trees and 24 pine trees. What is the ratio of ash to oak to pine in its simplest form?

Ratio of	Ash	:	Oak	:	Pine
=	18	:	12	:	24
=	3	:	2	:	4

> Simplify by dividing by 6 throughout.

Exercise 14.2

1 **a)** What proportion of the squares are red?
 b) What is the ratio of red squares to blue squares?
 c) What is the ratio of blue squares to red squares?
 d) Write the ratio of red : blue in its simplest form.
 e) Write the ratio of blue : red in its simplest form.

2 Write these ratios in their simplest form.
 a) $3:6$ **b)** $15:5$ **c)** $6:4$ **d)** $20:50$
 e) $4:20$ **f)** $16:12$ **g)** $10:25$ **h)** $28:8$

3 The following items were bought at the school snack bar.

Item	Number
Chocolate bars	35
Ice creams	20
Crisps	25
Total	80

 a) What proportion of the items were ice creams?
 b) What proportion of the items were crisps?
 c) What is the ratio of chocolate bars to crisps?
 d) What is the ratio of ice creams to crisps?
 e) Write your answers to parts **c)** and **d)** above in their simplest form.

4 A packet of scotch eggs had this list of
nutritional information.
 a) What is the proportion of fat in the scotch egg?
 b) What is the proportion of protein?
 c) What is the ratio of carbohydrate to fibre?
   Give your answer in its simplest form.
 d) What is the ratio of carbohydrate to fat? Give your answer in its simplest form.

Protein	12 g
Carbohydrate	16 g
Fat	20 g
Fibre	2 g

5 Use a set of 1–9 digit cards.
 a) Shuffle and deal out five cards.
 b) Write down the proportion of odd numbers in your group of five cards.
 c) Write down the ratio of even to odd numbers.
 Repeat parts **a)**, **b)**, and **c)** five more times.

6 The proportion of girls in a class is $\frac{17}{30}$.
 a) What is the proportion of boys?   **b)** What is the ratio of girls to boys?

7 Copy this chart to show the hours in a day from midnight through to midnight:

	12	1	2	3	4	5	6	7	8	9	10	11	12
a.m.													
p.m.													

Write 'S' for sleeping in each of the boxes during which you are usually asleep.
Write 'Sc' for hours you normally spend in school.
Write abbreviations for activities you do in the other hours, for example 'H' for
homework, 'O' for out, 'TV' for watching television, and so on.
 a) Write the ratio of hours spent sleeping to hours spent awake.
 b) Choose five other ratios to describe your day, for example the ratio of homework to
   time spent in school.

14.3 Ratio

Key words
ratio

⊕ Divide a quantity into two parts in a given ratio

To divide a quantity in a given **ratio**, there are four stages:

For example, to divide £40 in the ratio 3 : 7

a) Find the total number of parts by adding the two parts of the ratio: 10 parts (3 + 7).

b) Work out how much each part is worth: £40 divided by 10 is £4.

c) Multiply each part of the ratio by this amount: 3 × £4 = £12, 7 × £4 = £28.

d) Check that the two parts are in the given ratio, and that the total is correct.

  £12 : £28 = 3 : 7 (dividing by 4)

  £12 + £28 = £40

Example   Divide £36 in the ratio 5 : 4

```
Ratio                          Total
       5   :   4                 9
                                   ⌐
                                   │ ×4
                               36 ⌐┘
     5 × 4    4 × 4
       £20  :  £16
Check:  £20 : £16 = 5 : 4
        20 + 16 = £36 ✓
```

Exercise 14.3

1 a) Share 12 cards in the ratio 1 : 3

```
      Red  :  Black      Total
       1   :    3          4
                              ⌐
                              │ ×3
                          12 ⌐┘
```

b) Divide 28 green and blue balls in the ratio 5 : 2

```
      Green :  Blue      Total
       5   :    2          7
                              ⌐
                              │
                          28 ⌐┘
```

> Remember to check your answers, by adding up the shares to check if they sum to the total.

c) Share 40 black and white tiles in the ratio 3 : 5

```
      Black :  White     Total
       3   :    5
                              ⌐
                              │
                          40 ⌐┘
```

2 Divide £56 in the ratio 4 : 3

3 Divide 600 g in the ratio 7 : 3

4 Divide 189 cm in the ratio 4 : 5

5 Share £2400 in the following ratios:

  a) 1 : 5

  b) 3 : 1

  c) 5 : 3

  d) 3 : 7

6 Chelsea FC have won and lost their matches this season in the ratio 5 : 1.
If they played 36 games, how many games did they draw?

> Look back at the Example.

7 A 121 square metre garden contains lawn and patio in the ratio 5 : 6. How much is patio?

8 Angles on a straight line are in the ratio 2 : 7.
What is the size of each angle?

> Angles on a straight line = 180°.

9 Peter divides some postcards in the ratio 5 : 4 for pictures of cars and bikes.
He has 20 pictures of bikes. How many pictures of cars does he have?

<pre>
Cars   bikes
 5   :   4
             20
</pre>

> Find the 'multiplier'.

10 Jayne, Kim and Elaine bought a house.
They paid in the ratio 5 : 3 : 2.
If the house cost £75 000, how much did each pay?

Ratio			Total
5 :	3 :	2	10
5 × 7500	3 × ☐	2 × ☐	75 000
£37 500	?	?	

×7500

Solve it!

⊕ Identify the information needed to solve a problem
⊕ Be aware of a range of strategies

Problems can be presented in words or diagrams:
In each case the first thing we need to do is see which information is important, and which we don't need. Highlighting the important bits may help.

Here we look at different ways of solving problems. We call these different **strategies** . If one strategy doesn't work, try another. You will gradually get better at choosing the best strategy.

Here are three strategies you can try:

1) Make a guess and check the result.
2) Keep an organised list.
3) Make a table and look for a pattern.

Example 1 A book is lying open. The product of the two page numbers is 462. What are the page numbers?

A book is lying open. The **product** of the two page numbers is **462**. What are the page numbers?

20 × 20 = 400
20 × 21 = 420
22 × 23 = 506
21 × 22 = 462

Look at the units digits to help you.

Highlight the information you need.

Make a guess and check the result.
We need: two numbers that multiply to make 462 and are next to each other.

We know that 20 × 20 = 200 so try 20 and 21 (420 which is too small). Next try 22 × 23 (506 which is too big).

Example 2 Find all the factor pairs of 24. What is the biggest number you can make by adding each pair?

Find all the **factor pairs** of **24**.
What is the **biggest** number you can make by **adding** each pair?

Factors of 24	Factors add to give:
6 × 4	10
8 × 3	11
12 × 2	14
24 × 1	25

Highlight the important information.

Keep an organised list.
Guessing wouldn't be a sensible option here. Try several options and keep a table or list.

Exercise 14.4

1 Which information is not needed to answer the question?
   a) Fred and Albert both spend £5 every week on magazines, which each cost either £2 or £3. How much do Fred and Albert spend together in a whole year?
   b) Crisps cost 25p per packet and mints cost 35p. How many packets of crisps can I buy for £3?

Before you start the following questions, think about which strategy may be the best one to start with. Don't forget to highlight the important information!

2 Two brothers are three years apart. The product of their ages is 504. How old are they?

3 30 square tiles are used to make a frame. Find the greatest area that can be framed.

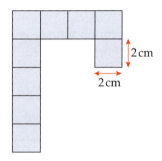

4 Find four consecutive odd numbers with a total of 80.

5 How many different ways can you share 240 sweets equally?

6 How many cans will there be in the tenth pile?

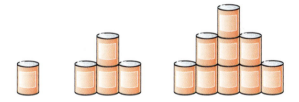

7 Pink and green triangular tiles are used to fill the hexagon, in a pattern that has two lines of symmetry. What are the different numbers of pink and green tiles that could be used?

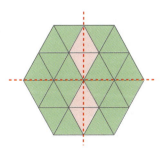

Multi-step problems

- Break complex calculations into simple steps
- Use working backwards as a strategy for solving problems

We have looked at problems in words and in diagrams.

Some problems are straightforward. Some problems are more complicated and need to be tackled in stages. These are called **multi-step** problems. It can sometimes be useful to work backwards from the answer that you need to find, in order to work out how to solve the problem.

Example 1   How much money is left over from £20 if I buy 6 comics at £2.50 each and 3 stickers at £1.25 each?

How much money is **left over** from **£20** if I buy **6** comics at

£2.50 each and **3** stickers at **£1.25** each?

$6 \times £2.50 = £15$ and $3 \times £1.25 = £3.75$

$£15 + £3.75 = £18.75$

$£20 - £18.75 = £1.25$

Highlight the important information.

The first step is to calculate the total costs of the comics and stickers.

Now add these together to see how much was spent.

Now find what's left from £20.

Example 2   Find the missing angle ∠ABC.

The angle we need is ∠ABC.

We can find ∠ABC using what we know.

If ∠DCE is 50° and ∠DCB is 65° then

∠BCA = 180 − 50 − 65

      = 65°

If ∠BCA is 65° and ∠CAB is 65° then

∠ABC = 180 − 65 − 65

      = 50°

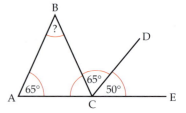

Angles on a straight line add up to 180°.

Angles in a triangle add up to 180°.

Exercise 14.5

1 A 50 cm square grid is made from wire.
If wire costs 40p per metre, find the total cost of wire
needed to make five grids.
Fill in the blanks to find the answer.

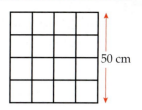

50 cm

	Calculation	Answer
The length of one grid	50 cm × 10	500 cm = ☐ m
The cost of one grid	☐ × 40p	☐
Cost of five grids	☐ × 5	£☐

2 The pie chart shows the lunch choices of a class.
Seven pupils have a school meal. How many go
home for lunch?
Fill in the blanks to find the answer.

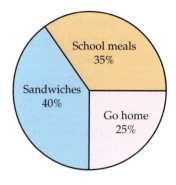

School meals 35%

Sandwiches 40%

Go home 25%

	Calculation	Answer
The number in the class altogether	7 people is 35% so 1 person is 35 ÷ 7 = 5%. If 1 person is 5%, the total number in the class is 100 ÷ 5 = ☐	☐
25% of the class	25% × ☐	☐

Can you find a quicker way of solving the problem?

3 Find the cost of making six picture frames if wooden
framing that is 5 cm wide costs £3.00 per metre.

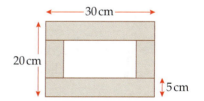

30 cm

20 cm

5 cm

4 Find the profit made by selling 20 books at £12.98 each
if they were bought for 50% of the selling price.

> Highlight the important information.

5 Two vertices of a square are at (3, 0) and (6, 2).
What are the coordinates of the other two vertices?

6 Find the area of this shape.

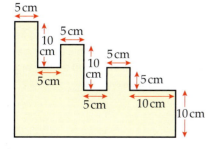

5 cm
10 cm
5 cm
5 cm
10 cm
5 cm
5 cm
5 cm
5 cm
10 cm
10 cm

Which strategy?

⊕ Choose and use appropriate and efficient operations and resources

A **strategy** is a way of doing something.

So far we have met these strategies for problem solving:

1) **Make a guess and check the result.**   2) **Keep an organised list.**
3) **Make a table and look for a pattern.**   4) **Work backwards.**

It always helps to highlight the important information.
Often there is more than one way to solve a problem.
Some ways are more **efficient** : they are shorter and clearer than other ways. How can we tell which is best to try first?

1) **Make a guess and check the result** works when you have good idea of what the answer is going to be and you know how to get closer to it. **Q1** is an example of this.

2) **Keeping an organised list** works when you need to keep a record of what you've done, perhaps to compare lots of different calculations, or when you need find the best answer. **Q2** is an example of this.

3) **Making a table and looking for a pattern** works when you are asked to find a general rule, or a certain term in a series. **Q3** is an example of this.

4) **Working backwards** is good when the problem has lots of steps and you need to work out what order to do them in. **Q4** is an example of this.

For **Q5–8**, you'll have to work out which strategy is best to use before you start to tackle the problem.

Example   16 squared tiles are used to make a rectangular picture frame.
What is the biggest area that can be framed?

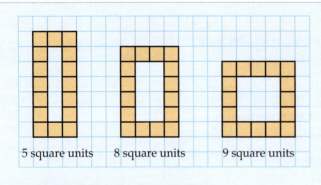

5 square units   8 square units   9 square units

Length	Width	Area
5	1	5
4	2	8
3	3	9

Draw diagrams and keep an organised list.
Guessing wouldn't be a sensible option here.
Draw several pictures and keep a table or list.
The answer is 9 square units.

Exercise 14.6

1 Two consecutive numbers multiply to give 650.
What are they?

2 Using the digits 3, 4, 5, 6 and 7, make a subtraction sum with the smallest possible answer.

3 How many squares are in the 10th pattern?

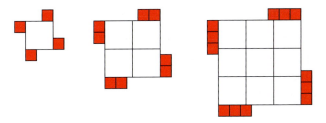

4 If the whole square has an area of 24 square units, find the area of the shaded region.

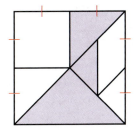

5 What is the biggest rectangular area I can make with 100 cm of wire?

6 A sports shop sells golf balls in different-sized packets:
Which packet is the best value?
If I wanted to buy 25 golf balls what would be the
cheapest way to buy them?

> 3 golf balls for £6.00
> 5 golf balls for £9.50
> 8 golf balls for £14.40

7 The mean of four numbers is 68. A fifth number is added which makes the mean 60.
What is the fifth number?

8 How many triangles are in the 20th pattern?

  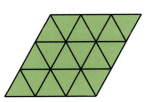

Drawing triangles 2

- Draw triangles, given two sides and the included angle, using a ruler and protractor
- Draw triangles given two angles and the included side, using a ruler and protractor

Key words
draw
sides
angles
included angle
included side
sketch

We can use a ruler and protractor to **draw** triangles accurately. To do this, we need information about their **sides** and **angles**.

If we know the lengths of **two sides** of the triangle and the size of the **included angle**, there is only one possible triangle that we can draw. We call this drawing a triangle using **SAS** (side, angle, side) information.

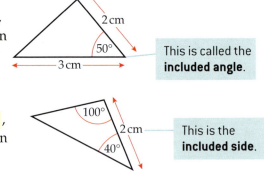

This is called the **included angle**.

If we know the size of **two angles** of the triangle and the length of the **included side**, there is only one possible triangle that we can draw. We call this drawing a triangle using **ASA** (angle, side, angle) information.

This is the **included side**.

It is often helpful to draw a **sketch** of the triangle, marking on all the information that we know, before we draw it accurately.

Example   Draw Triangle CDE where CD = 2.6 cm, $\angle$ECD = 45° and $\angle$EDC = 30°

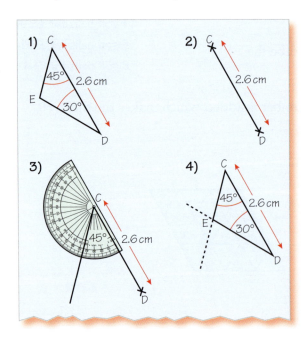

1) Draw a sketch first and mark on all the measurements.

2) Draw a base line CD of 2.6 cm. Mark a vertex for the angle at C.

3) Draw an angle of 45° with a long line.

4) Mark a vertex for an angle at D. Draw an angle of 30° and continue the line until it intersects with CE. The triangle is complete.

Exercise 15.1

1 Draw the shapes sketched below accurately. Measure the side marked with a star in each one and write down its length.

a)

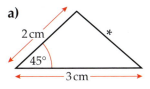

b)

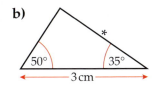

> Look back at lesson 2.4 pages 20–21.

c)

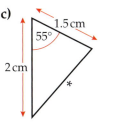

d)

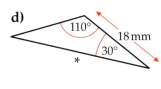

e)

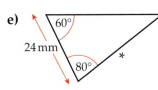

f)

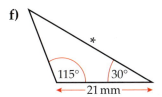

2 Draw a triangle with angles of 60° and 40°, where the side in between them measures 4 cm.
Remember to draw a sketch first.
Measure the third angle.
Add up the three angles.
Is your answer close to 180°?

3 Draw a triangle with angles of 53° and 64°, where the side in between them measures 4.5 cm.
Remember to draw a sketch first.
Measure and write down the lengths of the other two sides.

4 Draw triangle JKL where JK = 5cm, ∠JKL = 86° and ∠LJK = 22°.
Measure LK.

5 Draw triangle PQR where PQ = 5cm, ∠PQR = 100° and ∠RPQ = 20°.
Measure RQ.

6 Katie draws a quadrilateral PQRS.
PQ is 10 cm, ∠PQR is 90°, QR is 5 cm, ∠QRS is 110° and RS is 10 cm.
Draw a sketch of the quadrilateral PQRS. Now draw it accurately and measure the size of ∠RPS and SP.

15.2 Constructing triangles

◈ Construct a triangle given three sides, using a ruler and compasses

We can use a ruler and protractor or compasses to draw triangles accurately. To do this, we need information about their sides and angles.

If we know the lengths of **two sides** of the triangle and the size of the included angle , there is only one possible triangle that we can draw. We call this drawing a triangle using **SAS** information.

If we know the size of **two angles** of the triangle and the length of the included side , there is only one possible triangle that we can draw. We call this drawing a triangle using **ASA** information.

If we know the length of the three sides of the triangle, there is only one possible triangle that we can draw. We can construct the triangle using a ruler and compasses. We call this drawing a triangle using **SSS** information.

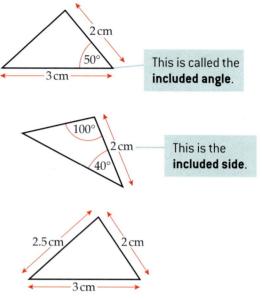

This is called the **included angle**.

This is the **included side**.

It is often helpful to draw a sketch of the triangle, marking on all the information that we know, before we draw it accurately.

Example Construct a triangle DEF with the following SSS information: DE = 2 cm, EF = 3 cm and FD = 4 cm

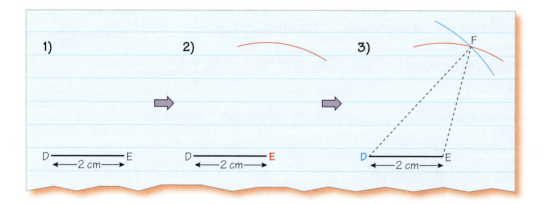

1) Draw a base line DE of 2 cm.
2) Open the compasses out to 3 cm and place the point at E. Draw an arc.
3) Open the compasses out to 4 cm and place the point at D. Draw an arc to intersect with the other arc. Join D and E to the point where the arcs intersect, F.

Exercise 15.2

1 Draw the shapes sketched below accurately:

a)

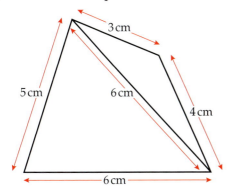

b)

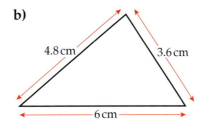

2 Construct triangles with the following SSS information:

    a) 4 cm, 3 cm, 6 cm   **b)** 5 cm, 6 cm, 7 cm   **c)** 3 cm, 4 cm, 5 cm.

3 Construct triangles with the following SSS information and then measure and write down the size of the three angles:

    a) 5 cm, 4 cm, 6 cm   **b)** 8 cm, 3 cm, 7 cm   **c)** 6 cm, 5 cm, 2 cm.

4 Construct this quadrilateral accurately:

5 Aaron is designing a tessellation of scalene triangles with sides of length 2 cm, 4 cm and 5 cm.

> A tessellation is a tiling pattern with no gaps.

    a) Draw six tessellating triangles with these dimensions.

    b) Will any scalene triangle tessellate? Can you explain why this is?

Investigation

6 **a)** What happens when you try to draw a triangle with sides 5 cm, 4 cm and 11 cm?

    b) Investigate other groups of three lengths that will not make triangles.

    c) What do you notice?

Shapes on coordinate grids

◈ Find coordinates of points determined by geometric information

Key words
coordinate grid
vertices
vertex

Rectangle
Four right angles.
Opposite sides are equal and parallel.

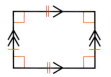

Square
A rectangle with four equal sides.

Parallelogram
Opposite angles are equal.
Opposite sides are equal and parallel.

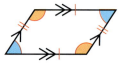

Kite
One pair of opposite equal angles.
Two pairs of adjacent equal sides.

Adjacent means next to.

We can plot all of these shapes on a coordinate grid . We can use what we know about the properties of these shapes to find the position of missing vertices .

For example, here three **vertices** of a rectangle are plotted on a coordinate grid.

We know a rectangle has four right angles and that the opposite sides are equal and parallel.

Using this information, the fourth **vertex** must be at the point (3, 5).

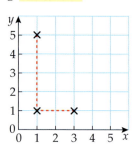

Example   Copy this diagram:

a) The points A, B and C are three vertices of a kite. Draw the kite and mark the fourth vertex D.

b) What are the coordinates of D?

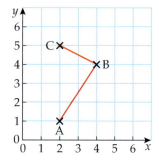

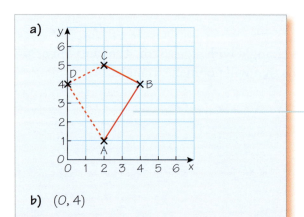

A kite has one pair of opposite equal angles and two pairs of adjacent equal sides. This means that CD must be the same length as CB and that the angles CDA and CBA must be equal.

b) (0, 4)

Exercise 15.3

1 Copy this diagram.

The points A, B and C are three vertices of a rectangle.

Complete the rectangle and mark the fourth vertex D.

What are the coordinates of D?

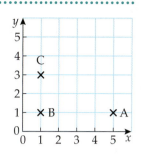

2 Each of these diagrams shows two sides of a quadrilateral.
For each one write down the coordinates of the fourth vertex.

a)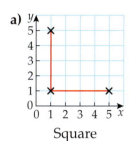

Square

b)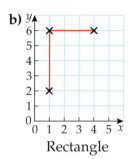

Rectangle

c)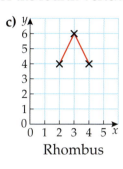

Rhombus

d)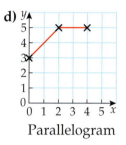

Parallelogram

3 Each of these diagrams shows three vertices of a parallelogram.
Write down the coordinates of the point that could be the fourth vertex.
There are three possible answers for each diagram. See if you can find all of them.

a)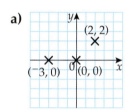

b)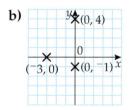

c)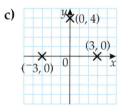

4 Plot the points: C(2, 1) and D(4, 3).
C and D are two vertices of a square.
Plot two more points to complete the square and mark them as E and F.
There is more than one possible answer. Find some other possible positions for E and F
and record their coordinates.

5 AB is one side of a rectangle which has an area of 12 square units.
Where could the other vertices be?

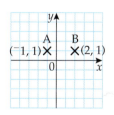

Investigation

6 Plot the following points: (2, 6) (4, 7) and (6, 6).
Mark another point to make a kite and write down the coordinates.
There is more than one possible answer. Find some other possible positions for the
fourth vertex and write down their coordinates.
What do you notice?

Mid-points

⊕ Find the mid-point of a line segment connecting two coordinates

If you know the **coordinates** of points A and B, you can find the **mid-point** of the **line segment** AB.

The **mid-point**, **M**, of this vertical line segment AB is halfway between the end points A and B.

Notice that:

- the x-coordinate of the mid-point is same as the x-coordinate of the end points A and B.

- the y-coordinate of the mid-point is the **mean** of the y-coordinates of the end points A and B. $\dfrac{(8 + 4)}{2} = 6$

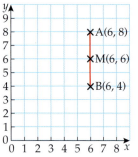

The **mid-point**, **M**, of this horizontal line segment AB is halfway between the end points A and B.

Notice that:

- the y-coordinate of the mid-point is same as the y-coordinate of the end points A and B.

- the x-coordinate of the mid-point is the **mean** of the x-coordinates of the end points A and B. $\dfrac{(3 + 7)}{2} = 5$

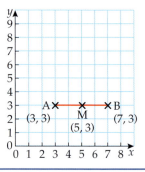

Example   Find the coordinates of the mid-points of the lines joining the following pairs of coordinates by plotting the points and drawing the lines.

a) (4, 2) and (4, ⁻4)   b) (3, 4) and (7, 4)

Check your results by calculation.

a)

The line is vertical so the x-coordinate of the mid-point is the same as the x-coordinate of the end points A and B.

The x-coordinate is 4.

The y-coordinate is $\dfrac{(2 + {}^-4)}{2}$

$= \dfrac{(2 - 4)}{2}$

$= \dfrac{{}^-2}{2}$

$= {}^-1$

The mid-point is at (4, ⁻1)

The line is horizontal so the y-coordinate of the mid-point is the same as the y-coordinate of the end points A and B.

b)

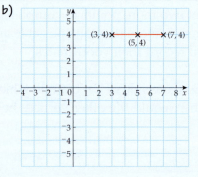

The x-coordinate is $\dfrac{(3 + 7)}{2}$

$= \dfrac{10}{2}$

$= 5$

The y-coordinate is 4.

The mid-point is at (5, 4)

Exercise 15.4

1 Find the coordinates of the mid-points of each of the following lines. Check your results by calculation.

a)

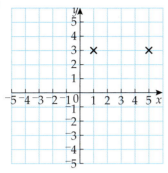

b)

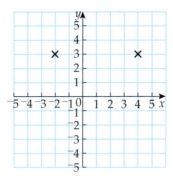

c)

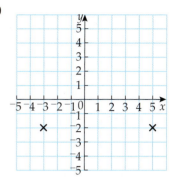

2 Find the coordinates of the mid-points of each of the following lines. Check your results by calculation.

a)

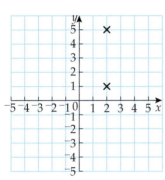

b)

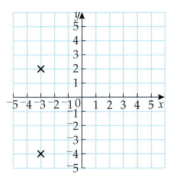

c)

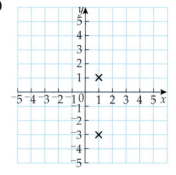

3 Find the coordinates of the mid-points of lines joining the following pairs of coordinates by plotting the points and drawing the lines:

a) (6, 2) and (6, 8)   **b)** (5, 3) and (9, 3)   **c)** (⁻2, 7) and (⁻2, ⁻1).

4 Here are some mid-points of lines drawn on a coordinate grid:

a) (4, 5)   **b)** (8, ⁻1)   **c)** (⁻5, 3)

The end-points of the line segments are all more than ⁻10 but less than ⁺10.
Write down some possible pairs of coordinates for the end-points of the line segments.

5 a) Look at this triangle.

Find the mid-point:
  i) of the side BC
  ii) of the side AC
  iii) of the side AB.

b) Look at the *x*-coordinate of the mid-point of the side AC and the *x*-coordinate of the mid-point of the side AB. What do you notice?

c) Look at the *y*-coordinate of the mid-point of the side BC and the *y*-coordinate of the mid-point of the side AB. What do you notice?

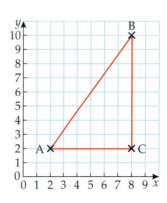

Shapes and paths

⊕ Use Logo to explore constructions of shapes and paths

We can use dynamic geometry programs such as LOGO to **construct** shapes.
To **construct** shapes, we use two different types of instructions:

movement in a **straight line** such as **forward 50 units**.

50

turning movements with a **direction** such as **right 60°**.

60°

We can use the fact that the **angles** on a **straight line**
add up to **180°** to help us calculate what turning
movement we need.

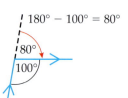
$180° - 100° = 80°$
80°
100°

Example 1 Write the instructions for drawing this path using LOGO:

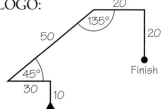

Fd 10   Lt 90   Fd 30   Rt 135   Fd 50   Rt 45   Fd 20

Rt 90   Fd 20

Example 2 Draw a path by following these instructions:

Rt 18   Fd 10   Lt 72   Fd 10   Lt 72   Fd 10   Lt 72   Fd 10   Lt 72   Fd 10

What shape have you drawn?

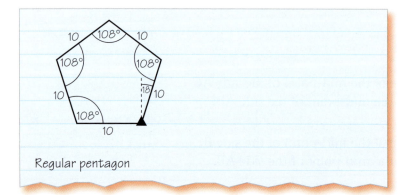

Regular pentagon

Exercise 15.5

1 Write the instructions for drawing these shapes using LOGO:

a)

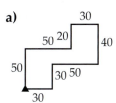

b)

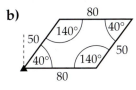

2 Write the instructions for drawing these paths using LOGO:

a)

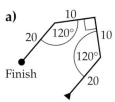

b)

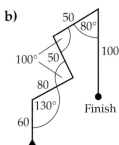

3 Draw a diagram with labels to represent the compass rose using LOGO:

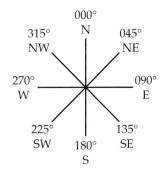

Investigation

4 A **unicursal** line is a line that stops and starts at the same place.
Using turns of 60° and 120° only, design three different unicursal lines.
How did you make sure that your line stopped and started in the
same place?

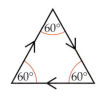

5 **a)** Draw paths, using paper or ICT, by following these instructions:
    i) Lt 30 Fd 60 Rt 60 Fd 60 Rt 60 Fd 60 Rt 60 Fd 60 Rt 60 Fd 60
        Rt 60 Fd 60
    ii) Lt 30 Repeat 6 [Fd 60 Rt 60]
  b) What do you notice?
  c) Investigate using the 'repeat' command to draw other regular polygons.

15.6 Visualising 3-D shapes

- ⊕ Use other 2-D shapes to visualise and describe 3-D shapes
- ⊕ Be able to draw 2-D representations of 3-D shapes

Key words
face
edge
vertex
cube
cuboid
tetrahedron
prism
pyramid
isometric

A **face** is the flat surface of a solid.

An **edge** is where two faces meet.

A **vertex** is where three or more edges meet.

A **cube** has six identical square faces.

A **cuboid** has three pairs of rectangular faces. Opposite faces are the same shape and size.

Here are some other types of **3-D** shape:

A regular **tetrahedron** has four equilateral triangles as faces.

A **prism** has two identical faces. It is the same shape all the way through.

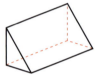

> You name a prism by the shape of its cross-section. This is a triangular prism.

A **pyramid** has triangular faces with a different shape for the base.

> You name a pyramid by the shape of its base. This is a square-based pyramid.

We can use **isometric** paper to draw representations of **3-D** shapes such as **cubes** and **cuboids**.

Example 1 Describe this shape.

> This shape is a hexagonal prism. It has two hexagonal-shaped faces and six rectangular faces. It has 18 edges and 12 vertices.

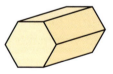

Example 2 Draw this shape on isometric paper.

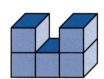

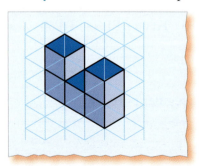

> Draw all of the front faces first, then the side faces and finally the top faces.

Exercise 15.6 ....

1 Write down the names of each of these shapes and record how many faces, edges and vertices they have.

a)   b) 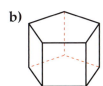   c)   d)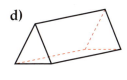

2 Use isometric paper to draw these shapes made from cubes. Shade the front faces.

a) 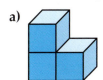   b) 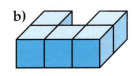   c) 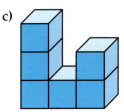   d)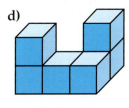

> Draw all the front faces first.

3 Here are three views of the same cube. Which coloured faces are opposite each other?

4 Imagine that you have two identical regular tetrahedra.
Place them together, matching face to face.
Draw a sketch of the new shape that is formed.
How many faces, edges and vertices does it have?

5 This is a photograph of one of the pyramids in Egypt.
Imagine that you are in an aeroplane, flying directly
over the top of it. Sketch an outline of what you would see.

Investigation

> Do not count rotations of the same model.

6 There is only one possible model that can be made from two cubes.
There are only two possible models that can be made from three cubes.
Make models from four and five cubes and draw them on isometric
paper. How can you be certain that you have made all the possible models?

Constructing nets 1

⊕ Identify nets for cubes and cuboids
⊕ Construct nets for cubes and cuboids

A cuboid has three pairs of rectangular faces.

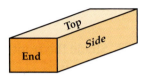

A cube has six identical square faces. The diagram shows a **net** for a **cube**.

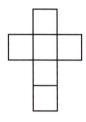

The diagram shows a net for a **cuboid**.

We can **construct** nets for cubes and cuboids.

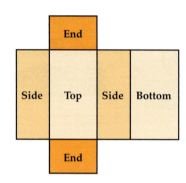

Example   Construct the net for this cuboid.

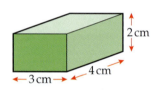

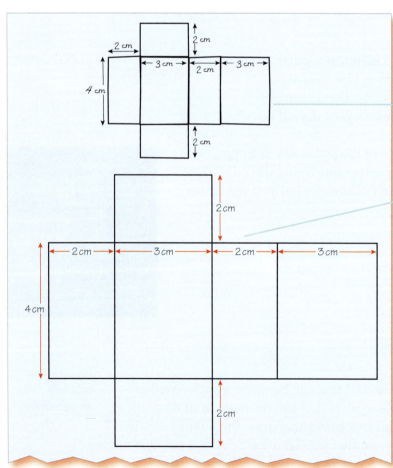

First, draw a sketch marking on the most important measurements.

Construct the net accurately using a sharp pencil and a ruler.

Exercise 15.7

1 Which of these are nets for a cube?

a)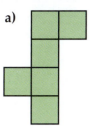

b)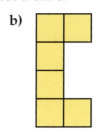

c) Draw them out on squared paper and make them to check.

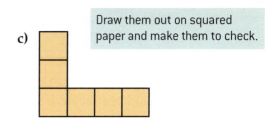

2 Construct the nets for these cuboids using centimetre-squared paper:

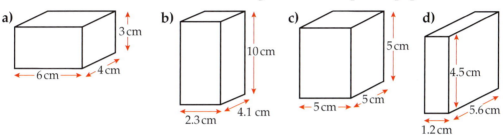

a) 3 cm, 6 cm, 4 cm

b) 10 cm, 2.3 cm, 4.1 cm

c) 5 cm, 5 cm, 5 cm

d) 4.5 cm, 5.6 cm, 1.2 cm

First sketch the nets with all the dimensions you know. This will also help you fit it on the page.

3 Which of these are nets for a cuboid? Draw a sketch of any 3-D cuboids that can be made.

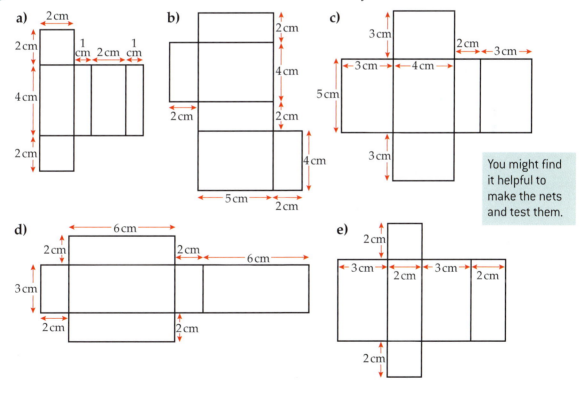

You might find it helpful to make the nets and test them.

4 A shoe box measures 10 cm by 10 cm by 30 cm. Sketch the net. Sketch the net for a box which is twice as wide, twice as long and twice as deep. Calculate the area of card needed to make the two nets and write down anything you notice.

Constructing nets 2

◈ Use a ruler and protractor to draw simple nets of 3-D shapes

A **net** folds up to make a 3-D shape. After sketching a net we can draw it accurately using a ruler and protractor, if we know enough facts about the 3-D shape we want to make.

Look back at page 192 for a reminder about nets for cubes and cuboids.
Here are some other examples of nets:

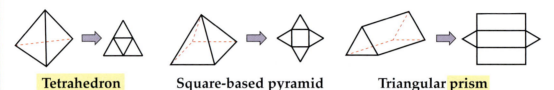

Tetrahedron   **Square-based pyramid**   **Triangular prism**

Example   Construct the net of a regular tetrahedron with sides of length 5 cm.

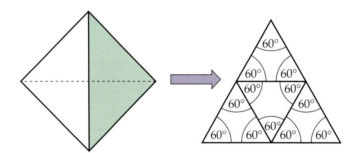

Sketch the net. We know that the faces are all equilateral triangles, so all the angles must be 60°. Mark on the measurements. Decide where to start. Use a set square and ruler to draw parallel lines wherever you can.

This is a suggested order of drawing but there are others.

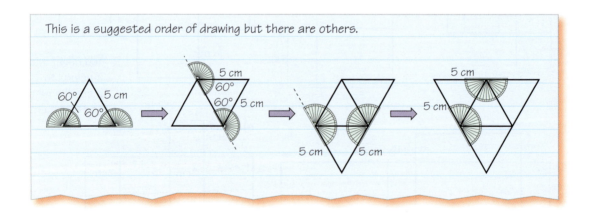

Exercise 15.8

1 What 3-D shapes will these nets make? Copy them accurately onto triangular dotty paper and assemble them.

a)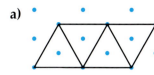

b)

2 Construct the net for this 3-D shape using square dotty paper.

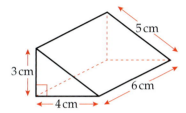

3 Construct the nets for these 3-D shapes using a ruler and protractor.

First sketch the nets with all the dimensions you know. This will also help you fit it on the page.

a)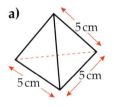

b)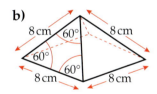

c)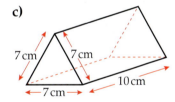

4 Design boxes for each of these new chocolate bars:

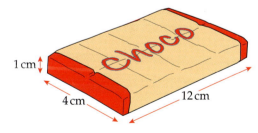

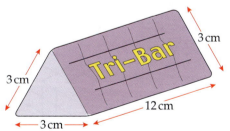

Volume

⊕ Calculate the volume of a cuboid
⊕ Solve problems involving cuboids

Key words
volume
cuboid
cross-section

The **volume** of a 3-D shape is the space contained inside it. We measure volume in cubes. Examples of units of volume are cm^3 and m^3.

The volume of a litre carton of milk is 1000 cm^3. The volume of a classroom is approximately 100 m^3.

You can think of the volume of a **cuboid** as cubes inside the shape.

Look at this cuboid:

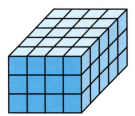

There are 4 × 3 = 12 cubes in each 'slice'.

The 'slice' is called the **cross section** of the cuboid.

There are 5 'slices', each containing 12 cubes, so the total number of cubes is 5 × 12 = 60 cubes

The volume of the cuboid is 60 cubic units.

The volume of any cuboid can be found by multiplying the width by the length and height.

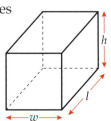

You can also think of the volume as the area of the cross section multiplied by the length.

Example   Work out the volume of this cuboid.

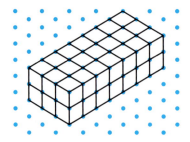

There are 6 cubes in each 'slice'.

There are 7 'slices' so the volume is 6 × 7 = 42 cubic units

First, count how many cubes there are in each 'slice'. Then count how many 'slices' there are.

Exercise 15.9

1 Count the cubes needed to make each of these cuboids.

a)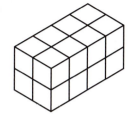

b)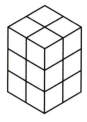

c)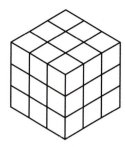

2 Calculate the volume of each of these cuboids.

a)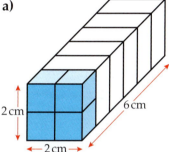

b)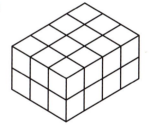

3 Work out:
    i) the volume   **ii)** the surface area of each of these cuboids.

a)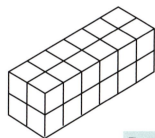

b)

    c) What do you notice about the volumes?
    d) What do you notice about the surface areas?

> The surface area is the total area of all the faces. You might find it helpful to draw out all the faces.

4 This cuboid has a volume of 36 cubes.
There are 9 cubes in each 'slice'.
How many 'slices' are there?

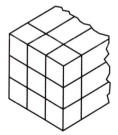

5 A cuboid has a volume of 24 cm³.
The diagram shows one possible cuboid.
Draw, on **isometric paper**, other possible cuboids with
volume 24 cm³. Make sure you label all the dimensions clearly.

Planning is the first part of the data handling cycle:

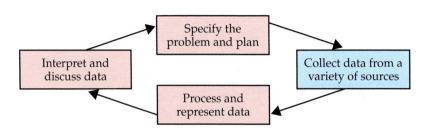

You can collect information from primary or secondary sources. A **primary source** is data you collect yourself. A **secondary source** is data that has already been collected. You can find secondary data from sources such as the Internet, a book or a CD-ROM.

You will also need to decide on **sample size** (the number of people you will include). You need a sample size that is large enough to give reliable data, but not so large that it takes too long to collect and process the data.

Example A teacher wants to find out if the school canteen should sell a greater variety of food.

a) What questions could be asked? Design tick boxes for them.

b) Who should be asked?

c) How many pupils should be asked?

a) How often do you use the canteen each month?

Number of visits:

0–4	5–9	10–14	15–19	20+

How much do you spend each visit?

Up to 50p	51p–£1	£1.01–£1.50	£1.51–£2	More than £2

What other types of food should the canteen sell?

pasta	soup	cakes	other

Include an 'other' or 'more than' box so you cover all possible replies. In addition to tick boxes, you might also like to leave space for pupils' own suggestions.

b) Pupils of different ages, both boys and girls, not just pupils who are in the canteen at the time.

c) She should ask about 40–50 pupils.

Asking too few will not give enough information. Asking too many will take too long.

Exercise 16.1

1 Write down how you would collect data for the following. Choose from a **survey**, an **experiment** or a **secondary source**.
   a) The number of times a dice is rolled until a score of 30 or more is reached.
   b) 'Can tall people run faster than shorter people?'
   c) The most popular film among pupils in your school.
   d) The most popular film this year in the United Kingdom.
   e) The most common bird seen in the school grounds.
   f) 'Do we need a youth club at school?'
   g) The populations of Britain's ten largest cities.

2 Neil has made a questionnaire about how often pupils play sport. He stands outside the school library at break, and asks six Year 7 girls and three Year 9 boys to fill in the questionnaire.

> Name: Age:
> Class: Boy/Girl:
> How often do you play sport?
> Often ☐   Sometimes ☐   Never ☐   Depends ☐
> What sports do you play?
> Football ☐   Swimming ☐   Tennis ☐
> Children don't play enough sports, do they?   Yes ☐   No ☐

   a) Show how his three questions can be improved.
   b) Write down at least three mistakes Neil has made when carrying out his questionnaire.
   c) Explain why Neil wants to collect information such as age, name and class.

3 Design data collection sheets for the following:
   a) The times taken by the top male and female sprinters to run 100 m
   b) The type of vehicle passing the school gates at different times of the day
   c) The number of *heads* showing when three coins are thrown.

> Decide how many times you would carry out this experiment.

4 Design 'tick boxes' for these questions:
   a) How many times were you late for school last week?
   b) What type of music do you like?
   c) How many books do you read in a month?
   d) How much do you spend on clothes in a month?

> Do your tick boxes cover **all** possible answers?

> Use intervals or groups for your answers to parts **c)** and **d)**.

5 'Between 6 pm and 10 pm, children spend more time playing computer games than any other activity.' Design a questionnaire to find out if this statement is true.

Investigation

6 Plan an investigation to test one of the following:

   – Compared with boys of the same age, teenage girls get more pocket money.
   – The taller you are, the bigger shoe size you take.
   – Compared with biscuits, crisps contain more fat.

   Include a question that involves number and put your data into groups.

Collecting data

⊕ Collect data using a suitable method

In Lesson 16.1 we saw how to plan an investigation. In this lesson, we will focus on methods of collecting data.

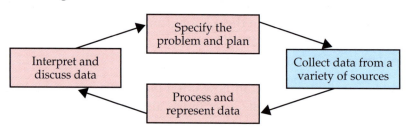

It is important to design a suitable **data collection sheet** that will allow you to record a range of data and process this data quickly. You will also need to decide how much data to collect – this is your **sample size** .

Example   A local business wants to find the best location to open a new sportswear shop.
   a) Design a data collection sheet to find out:
      i) Shopping preferences for different age groups
      ii) Why people use certain shops
      iii) How often they buy sportswear.
   b) How many people should be questioned?

a) i)

Location \ Age	16–25	26–35	36–45	46–55	56–65	66+
High street/local shops						
Out-of-town supermarket						
Internet						

Age categories should not overlap or have gaps.

ii)   What affects your choice of shop? (Tick 1 box)

   Convenience ☐   Cheap prices ☐   Transport ☐

   Large choice ☐   Other ☐

Giving a small number of choices makes analysing the data easier.
Include an 'other' box.

iii)   How often do you buy sportswear?

   More than once a month ☐

   Once a month ☐

   Less than once a month ☐

Categories should cover all possible answers.

b)   Between 50 and 100 would be about right.

Asking a small number, for example, 20 people, would not give enough data. Asking too many would take too long.

Exercise 16.2

1 'I put cold water out for birds in the winter because hot water freezes faster than cold.'
Plan an experiment to test this statement. Design a data collection sheet, and describe how you would collect the data.

2 Produce a plan to compare the performance of two football, cricket or rugby teams or two athletes. Which factors will you compare?
Design a data collection sheet.
How many years of data will you use?

3 Plan a survey to find the answer to this question: 'Does the time of day affect the amount or types of traffic on the road?'
Design data collection sheets to record:
- The amount of traffic passing a point in a minute
- The different types of vehicles
- The number of passengers in a vehicle.

4 'The frequencies of each letter used in the English language, from greatest to least are:
   ETOAN   IRSHD   LCWUM   FYGPB   VKXQJZ'.

Design an experiment to test some or all of this statement.
Include a data collection sheet.
Where will you collect your data?
Will you use different sources?
How much data will you collect?

5 How long does it take to say the alphabet backwards?
Plan an investigation and design a data collection sheet.
How many times will you carry out this experiment?

6 Design a questionnaire to find out about the eating habits of pupils in your school.
Decide what type of information to collect and design a data collection sheet.
How many pupils will you ask?
Will you ask pupils of different gender and from different year groups?
Where will you carry out your survey?

7 Design an experiment to test if people can tell butter from different types of margarines and spreads. Design a data collection sheet.
How many people will you ask?
What different types of people will you ask?
Where will you carry out your survey?
What special equipment will you need?

Investigation

8 Carry out one or more of the above investigations, recording the information on your data collection sheet.

Displaying data

◉ Choose a suitable diagram to represent data

When you have collected the data you need, you can draw graphs and diagrams. Choose an appropriate diagram for the type of data. For example:

- Use a **pie chart** to show how something is shared or divided
- Use a **bar chart** or **bar-line graph** to show the numbers in each category.

Example   A bus company collects data about a bus service to the town centre:

Reason passengers use the service	No car	Bus is cheap	Good for the environment	Difficult to park in town
Frequency	15	11	11	4

Number of passengers using the service	0–9	10–19	20–29	30–39
Frequency	16	31	20	26

Number of times a passenger uses this bus service each week	1	2	3	4	5
Frequency	3	8	10	18	11

Draw a suitable diagram for each table of results.

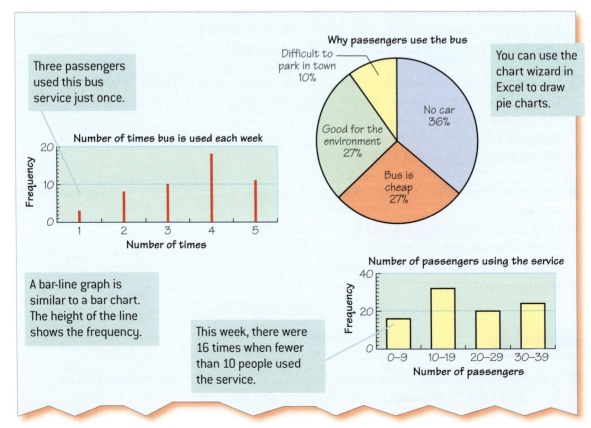

Three passengers used this bus service just once.

A bar-line graph is similar to a bar chart. The height of the line shows the frequency.

This week, there were 16 times when fewer than 10 people used the service.

You can use the chart wizard in Excel to draw pie charts.

Exercise 16.3 ....

① Helen collects some data about the top 40 singles music chart.

Number of weeks in the chart	1–10	11–20	21–30	31–40	41–50
Number of singles	28	6	3	2	1

A bar chart would be suitable for this data.

Type of music	R and B	Dance	Pop	Other
Number of singles	6	18	12	4

There is more about pie charts in lesson 16.5 pages 206–207.

Draw suitable diagrams for her information.

② Joaquin collects some data about the number of words in a sentence. He chooses a book written for a child and a book written for an adult.

Number of words in a sentence	1–5	6–10	11–15	16–20	21–25	26–30
Book A	2	15	18	17	26	22
Book B	7	48	29	15	1	0

a) Draw a suitable diagram for each book.
b) Which book was written for a child? Explain your answer.

③ Write down which type of diagram you would choose to represent each of these tables. If you have time, draw the diagrams.
a) A school collects data about pupils being late for school.

Number of times late in a year	0–4	5–9	10–14	15–19	20–24
Number of pupils	55	24	8	2	1

Reason late	Bus late	Overslept	Missed bus	Other
Number of times	135	90	30	45

b) Melissa conducts a survey about part-time work and pupils at her school.

What type of work is your part-time job?

Type of work	Shop	Paper-round	Café/restaurant	Babysitting	Other
Frequency	14	6	10	6	4

How many hours do you work each week?

Number of hours	1	2	3	4	5
Frequency	16	10	12	0	2

Investigation

④ Draw suitable diagrams using the results from the investigation you carried out in the previous lesson.

Line graphs

⊕ Draw and interpret line graphs

You can plot some types of data on a **line graph** .

Start by marking out the **horizontal** axis in equal steps. (It is a good idea to do this in pencil first.) Check that your lowest and highest values will fit on the line. If they will not fit, you will need to change the size of your steps. Repeat for the **vertical** axis, then plot the points. Join the points with a line. This line can be used to estimate unknown values.

Example 1 Draw a graph showing the number of cars owned in Great Britain to the nearest million.

Year	1978	1980	1982	1984	1986	1988	1990	1992	1994	1996	1998	2000
Number of cars (millions)	14	15	15	16	17	18	20	20	20	21	22	23

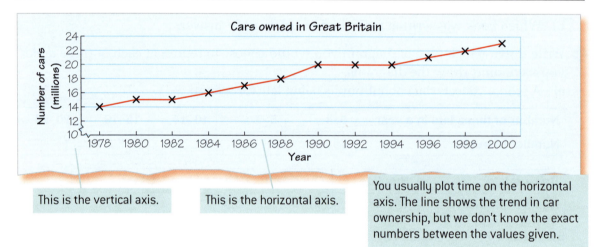

This is the vertical axis.　　This is the horizontal axis.

You usually plot time on the horizontal axis. The line shows the trend in car ownership, but we don't know the exact numbers between the values given.

Example 2 This graph shows pairs of numbers that multiply together to give 48.

For example $24 \times 2 = 2 \times 24 = 48$

Use the graph to find the missing numbers to make 48 each time.

a) $12 \times \square$

b) $8 \times \square$

c) $32 \times \square$

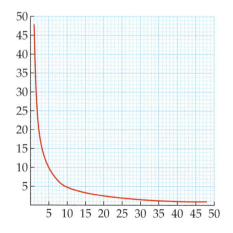

a) 12×4

b) 8×6

c) 32×1.5

Exercise 16.4

1 Draw a time series graph to show the number of televisions owned in Great Britain. Between which five years was the increase greatest? Use your graph to help you.

Year	1960	1965	1970	1975	1980	1985	1990	1995	2000
Number of TVs (millions)	16	17	18	19	20	21	22	23	25

2 Draw a time series graph to show the value of a car against its age. Between which two years does the value drop the most?

Age of car (years)	0	1	2	3	4	5
Value (£)	12 900	8300	7110	6000	5000	4100

3 This graph shows pairs of numbers that multiply together to make 60. Use the graph to find these missing numbers.

a) $30 \times \square$

b) $4 \times \square$

c) $8 \times \square$

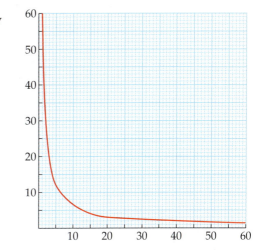

4 Brian is on holiday in Malaysia. To change £ to Malaysian $ he multiplies the number of £ by 6.

£	2	4	6	8	10	12	
Malaysian $	12				60		

a) Copy and complete the table.

b) Draw a graph to change £ to Malaysian $.

Use your graph.

c) £3 is approximately how many Malaysian $?

d) Malaysian $30 is approximately how many £?

e) £5.50 is approximately how many Malaysian $?

5 The table shows the temperature of a beaker of warm water as it cooled.

Time (minutes)	0	1	2	3	4	5	6
Temperature °C	62	59	56	54	52	50	49

a) Draw a graph for this data.

b) Estimate at what time the temperature was 53 °C?

c) Estimate the temperature at 1.5 mins.

Plot time along the horizontal axis.

Use Q4 to help you.

Investigation

6 Use a newspaper or the Internet to find the exchange rate for the money in another country. Draw a graph to show the exchange rate for amounts between 0 and £200.

More about pie charts

⊕ Draw and interpret pie charts using ICT

> **Pie charts** are very useful to show **proportions** of amounts (how something is shared or divided).

Example 1 Shyam calculates how she spends her time away from home during a school day.

Travel	Lessons	Lunch break	After school clubs
10%	55%	15%	20%

Draw a pie chart to represent this information.

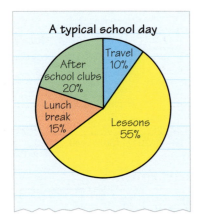

Complete the cells in Excel to show this information.

Use the chart wizard and select *Pie*.

Select the *chart sub-type* that shows a 2-D pie chart.

Remember to fill in the *chart title* box with a suitable title.

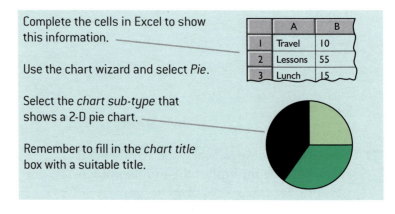

	A	B
1	Travel	10
2	Lessons	55
3	Lunch	15

Example 2 Mike asks pupils at his school 'Should we have school uniform?' Here are the replies:

Answer	Yes	No	Don't know
Number of pupils	35	10	5

Draw a pie chart to represent his information.

Answer	Yes	No	Don't know
Number of pupils	35	10	5
Percentage of total	$\frac{35}{50} = \frac{70}{100} = 70\%$	$\frac{10}{50} = \frac{20}{100} = 20\%$	$\frac{5}{50} = \frac{10}{100} = 10\%$

Should we have school uniform?

Don't know 10%
No 20%
Yes 70%

Exercise 16.5 ···

1 Helen spends her allowance on different items. Draw a pie chart to represent this information.

Clothes	Sport	Going out	Save
30%	40%	20%	10%

If Helen receives £10 each month, how much money does she save?

2 Draw a pie chart showing the population of Great Britain, by age group.

Age group	0–14	15–64	65+
%	20%	65%	15%

3 Phil asks 40 people what they ate for breakfast.
Draw a pie chart to represent this information.

Type of breakfast	Number of pupils
Cereal	16
Toast	10
Cooked	6
Other	8

4 200 people are asked their age and the question:

'How often do you recycle the carrier bags from your shopping?'

Response / Age	Always	Often	Rarely	Never
Under 30	25	40	20	15
Over 30	32	44	5	19

Draw a pie chart for each age group for this data. Write a sentence to sum up what the pie charts show.

5 Draw a pie chart to show the favourite 'take-away' for a group of friends.

Indian	Chinese	Pizza	Burgers	Fish & chips
4	3	5	2	6

6 A P.E. teacher draws pie charts to show the results of the football and netball teams. Can you tell from the charts if the teacher is right? If not, suggest what other information you would need.

The netball team won half as many games as the football team. The football team won 10 games so the netball team must have won 5.

Football

Netball

7 The school librarian asks some children about their favourite type of book.

Draw a pie chart for her data.

Type of book	Humour	Science-fiction	Romance	Thriller	Non-fiction
Number of pupils	15	10	7	6	12

Drawing and interpreting diagrams

- Draw and interpret bar charts and pie charts
- Choose the most suitable diagram to represent data

When drawing diagrams it is important to choose the most suitable one to represent the data. For example, **bar charts** are good for showing categories, and **pie charts** are good for showing how a whole is divided up. **Compound bar charts** are useful when we want to display the data for two or more groups (for example boys and girls).

Example Represent this data with a diagram.

Favourite hobby	Sport	Music	Collecting	Other
Number of girls	12	8	2	3
Number of boys	15	3	8	6

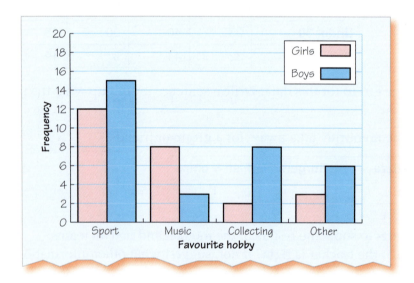

Exercise 16.6

1 The number of pupils with a mobile phone at a school are shown below.
Draw a suitable diagram for this information.

Year	7	8	9	10	11
Number of girls	29	57	63	82	80
Number of boys	20	41	61	90	91

Look at the Example.

2 Neil collects data about football matches played by his local teams at the weekend.

Outcome	Win	Draw	Lose
Number of games	5	1	4

Number of goals scored	0	1	2	3	4
Number of teams	3	4	2	0	1

Team	Albion Utd	Hot shots	City	Hurricanes	Queenstown
Attendance figure	2300	1900	2900	1500	800

Draw a suitable diagram for each set of data.
Explain why you chose that particular diagram.

3 The table shows the number of medals won by the Netherlands in the 2000 Olympic games. Draw a suitable diagram to represent this data.

Type of medal	Gold	Silver	Bronze
Frequency	12	9	4

A Gold medal scores 3 points, a Silver medal scores 2 points and a Bronze medal scores 1 point. How many points did the Netherlands score in total?

4 The average maximum temperature for each month in Cornwall is shown below. Draw a suitable diagram for this information.

Month	J	F	M	A	M	J	J	A	S	O	N	D
Maximum temperature (°C)	8.3	8.0	9.4	11.5	14.1	16.8	18.6	18.7	17.1	14.4	10.9	9.2

a) Which month was the warmest? Which was the coldest?
b) What was the difference in the temperature between these two months?

5 Nai has drawn a pie chart to show the population of some London boroughs.
a) Explain why this is not a good choice of diagram for this data.
b) Draw a more suitable diagram.

London borough	Population
Hackney	200 000
Islington	175 000
Lewisham	250 000
Kensington & Chelsea	160 000

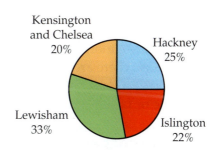

6 Draw a suitable graph to show the populations of these countries.

Finland	5 million
Luxembourg	$\frac{1}{2}$ million
Netherlands	16 million
Portugal	10 million
Sweden	9 million
Switzerland	$7\frac{1}{2}$ million

> Using a scale of 2 squares for 1 million people may be useful.

Communicating results

✦ Interpret tables and diagrams
✦ Communicate the results of an investigation

When you have calculated **statistics** (for example the mean, mode, median and range) and drawn diagrams, you need to interpret your results, and answer the original question. You may need to write a short report.

Example 1

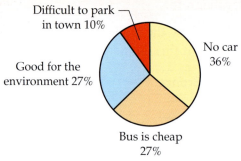

Why passengers use the bus

Difficult to park in town 10%
No car 36%
Good for the environment 27%
Bus is cheap 27%

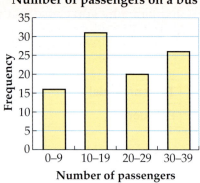

Number of passengers on a bus

Use these diagrams to decide if these statements are true, false or unknown:

a) The mode is the only average you can find from the pie chart.

b) The range of the number of passengers on the bus is $39 - 0 = 39$.

c) The modal class of the number of passengers is 30–39.

---

a) True

b) Don't know.

c) False. It is 10–19 passengers.

The **range** is largest number − smallest number. The data is grouped, so we do not know if the 30–39 class included the number 39, or if the 0–9 class included the number 0.

The **modal class** is the class with the highest frequency.

Example 2 'There are more boys in class 8B because there are 60% compared to 50% in class 8A'
True or false?

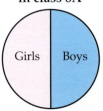

Boys and girls in class 8A

Girls | Boys

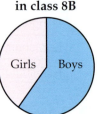

Boys and girls in class 8B

Girls | Boys

Don't know the answer. We cannot tell this from the information given.

The percentage of boys in 8B is larger, but we do not know how many pupils are in each class, so we can only compare percentages.

For example, if there are 30 pupils in 8A, then 50% = 15 boys. If there are 25 pupils in 8B then 60% is also 15 boys $\left(\frac{60}{100} \times 25\right)$. The number of boys would be the same.

Exercise 16.7

1 Member's of Ian's class draw a diagram to show the number of pages in their reading books.

Write down whether each of these statements is true or false, or if you need more information. Explain your answers.

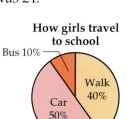

Number of pages in a reading book

a) The range of the number of pages is 249.

b) The modal class for number of pages is 100–149.

c) The number of books in the survey was 24.

2 Nikki collects information about how pupils travel to school.

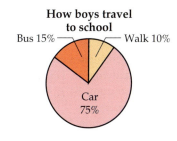

a) Nikki says her pie charts show that more boys than girls travel to school by car. Is Nikki correct?

b) Nikki collected information from 30 girls and 20 boys. Use this new information to write a short report about how children in Nikki's survey travel to school.

3 Fazan buys some peas still in their pods. He counts the number of peas in each pod and records them in a table.

For each of the following sentences say whether they are true, false or if you need more information.

Number of peas	Tally
0–2	ⵍⵍ ‖‖
3–5	ⵍⵍ ⵍⵍ ‖
6–8	‖‖
9–11	‖

a) He opened 24 pods in total.

b) Half of the pods had between 3–5 peas in them.

c) Some pods had no peas in them.

d) The last pod he opened had the most peas in it.

e) The modal class is 9–11 peas.

4 A group of children are comparing two different makes of mints. They each buy a bag containing 100 grams.

Mintees

Number of sweets	12–14	15–17	18–20	21–23	24–26
Number of bags	2	0	6	5	1

Mintburst

Number of sweets	12–14	15–17	18–20	21–23	24–26
Number of bags	1	2	4	7	1

a) Which is the modal class for Mintees?

b) Which is the modal class for Mintburst?

c) Adam says that the range of sweets in a Mintees pack is 26 − 12 = 14 sweets
Luke says that the range is only 10 sweets. Explain why Luke or Adam could be correct. Write a report to sum up which pack of mints is likely to provide you with more sweets per pack.

Index

acute angles 16, 18
addition
    decimals 92
    fractions 44
    integers 2
    inverse 58
    standard written method 92
algebraic expressions
    collecting like terms 50, 118
    like terms 50
    relationships of unknowns 116
    simplifying 50, 118, 118
    substitution 54
    terms 50
angle of rotation 106
angle sum, triangle 14
angles
    acute 16, 18
    exterior 17
    full turn 14
    half turn 14
    included 20
    measuring 18
    obtuse 16, 18
    protractor 16, 18
    reflex 16, 18
    right 16, 18
    sum at a point 14
    sum on a straight line 14
    triangles 14, 24
    vertex 20
    vertically opposite 22, 24
area
    metric units 66
    rectangle 66, 68
    square 140
    triangles 70
ascending sequence 10
averages 130, 134, 138

balancing equations 56, 58
balancing method 120, 122
bar chart 202, 208
bar-line graph 202
bias 36
brackets
    calculator 142
    expanding 52
    simplifying 52

calculator
    brackets 142
    decimals and fractions 40
    order of operations 142
    square numbers 140
    square root 140
capacity 86
Celsius 161
centre of enlargement 114
centre of rotation 106
cl (centilitre) 86
class intervals 136
cm (centimetre) 86
cm^2 (square centimetre) 66
collecting like terms 50, 118
common denominator 42, 44
common factors 100
common multiples 98
communicating results 210
compound bar chart 208
consecutive numbers 123
consecutive terms 10, 12, 156
conversion graphs 160
coordinate grid
    origin 78, 160
    outline 78
    shapes 183
    x-axis 78
    y-axis 78, 82
coordinates
    graphs 80
    mid-point of line 186
    x-coordinate 78
    y-coordinate 78
cross section 196
cube
    net 192
    surface area 72
    3-D shape 190
    volume 196
cuboid
    cross section 196
    net 72, 192
    surface area 72
    3-D shape 190
    volume 196

data
    class intervals 136
    collection 198, 200

comparison 134
display 202, 204
frequency tables 136
grouping 136
handling cycle 198
modal class 136
modal group 136
presentation 202, 204
primary source 198
sample size 198, 200
secondary source 198
statistics 134, 138
two-way tables 128
data collection
    frequency table 32
    sheet 200
    tallies 32
decimal point 62
decimals
    addition 92
    division 62, 146
    fractions 40
    multiplication 62
    ordering 88
    probability 28
    rounding 90
    subtraction 92
denominator 38, 40, 42
descending sequence 10
direct proportion 168
distributions, comparison 134
divisibility tests 6
division
    by 10, 100, 1000.... 62
    decimals 62, 146
    equations 120
    estimate by rounding 144
    remainder 144
    standard written method 144
    tests for divisibility 6
divisions marked on scales 64
divisor 144
drawing triangles 20, 180, 182

edge 190
efficient strategies 178
enlargement 114
equations
    balancing methods 56, 58

divisions 120
graphs 80
inverse operations 56, 58
multiples of x 58
simplifying 118
solving 56, 58
solving with algebra 122
straight-line graphs 78, 80, 82, 84
equilateral triangle 24
equivalent fractions 38, 40, 42, 44
equivalent point 104
estimating probability 34
event
frequency 130, 132
possible outcomes 30
probability 26
expanding the brackets 52
experiment 198, 210
experimental probability 36
exterior angles 17

face 190
factor pair 148
factor pairs 6, 100
factorisation 148
factors 100
Fahrenheit 161
formulae
constructing 126
definition 60
in words 124
relationships between variables 124
substitution 60
using 126
fractions
addition 44
cancelling 38
common denominator 42, 44
decimals 40
denominator 38, 40, 42
equivalent 38, 40, 42, 44
improper 44
lowest terms 38
mixed number 44
multiplication 46
numerator 38, 40
of amounts 46
ordering 42
probability 28
simplest form 38
simplifying 38
subtraction 44
frequency 130, 132
frequency tables 32, 136

g (gram) 86
general term 152, 154, 156
gradient 164, 166
gradient of a line 84
graphs
conversion 160
coordinates 80
drawing 162
equations 80
gradient 164, 166
horizontal axis 204
interpreting 164
plotting 80
real life 162, 164, 166
relationships between variables 164
slope 164, 166
straight-line 78, 80, 82, 84
vertical axis 204
y-intercept 82, 164
grouping data 136

HCF (highest common factor) 100
highest common factor (HCF) 100
horizontal axis 204
hundredths 48

image and object 104, 110, 112
improper fractions 44
included angle 20, 180, 182
included side 180, 182
indices 142
integers 2, 4
interpreting results 210
inverse
addition 4, 58
square number 140
square root 140
subtraction 4, 58
inverse operations 120, 122
solving equations 56, 58
investigation 198, 210
isometric paper 190
isosceles triangle 24

kg (kilogram) 86
kite 183
km (kilometre) 86
km² (square kilometre) 66

ℓ (litre) 86
LCM (lowest common multiple) 98
length 64, 86
like terms 50
line graph 204
line of symmetry 108

line segment 18, 186
LOGO program 188
lowest common multiple (LCM) 98

m (metre) 86
m² (square metre) 66
mapping diagram 74, 76
mass 64, 86
mean 130, 132, 134, 138
measuring instruments 64
median 130, 134, 138
mental arithmetic 148, 150
metric units conversion 86
mid-point of line 186
mirror line 104, 108
mixed number 44
ml (millilitre) 86
mm (millimetre) 86
mm² (square millimetre) 66
modal class 136, 210
modal group 136
mode 130, 134, 138
multi-step problems 176
multiples 98
multiplication
by 10, 100, 1000.... 62
decimals 62
estimating by rounding 94, 96
fractions 46
grid 52
standard methods 94, 96
multiplication table 98

negative whole numbers 2, 4
net
cube 192
cuboid 72, 192
square-based pyramid 194
tetrahedron 194
triangular prism 194
notation
centilitre (cl) 86
centimetre (cm) 86
degrees (angles) ° 14
degrees (temperature) °C & °F 161
enlargement A′ B′ C′ 114
gram (g) 86
kilogram (kg) 86
kilometre (km) 86
lines of equal length 23
litre (ℓ) 86
metre (m) 86
millilitre (ml) 86
millimetre (mm) 86
nth term 152, 154, 156
square centimetre (cm²) 66

square kilometre (km²) 66
square metre (m²) 66
square millimetre (mm²) 66
number line 2, 4, 88
numerator 38, 40

object and image 104, 110, 112
obtuse angles 16, 18
order of operations 142
order of rotation symmetry 108
origin 78, 160
outcomes of event 30

parallel lines 22, 84
parallelogram 183
partitioning 150
patterns and sequences 8, 154
percentage and probability 28
percentages 48
perimeter, rectangles 68
perpendicular bisector 104
perpendicular height 70
perpendicular lines 22
pie chart 202, 206, 208
place value grid 62
plotting graphs 80
polygons, symmetry 108
positive whole numbers 2, 4
powers 142
primary source 198
prime numbers 102
prism 190
probability
    bias 36
    comparing 36
    decimal 28
    estimating by experiment 34
    event 26
    event outcomes 30
    experimental 36
    fraction 28
    percentage 28
    random choice 28
    random event 26
    sample space diagram 30
    theoretical 28, 36
    using numbers 28
    using words 26
    what not to write 28
probability scale 26
problem solving
    formulae 126
    information needed 174
    multi-step 176
    strategies 174, 176, 178
proportion 168, 170, 206
protractor 16, 18, 180

pyramid 190

questionnaire 198, 210

random choice 28
random event 26
range 130, 134, 210
Rangoli patterns 105
ratio 170, 172
real life graphs 162, 164, 166
rectangle
    area 66, 68
    coordinate grid 183
    perimeter 68
reflection 104, 108
reflex angles 16, 18
remainder 144
right angles 16, 18
right-angled triangle 24
rotation 106, 108
rounding
    down 90
    estimating for division 144, 146
    to estimate multiplication 94, 96
    to nearest 10, 100, 1000.... 90
    to nearest hundredth 90, 146
    to nearest tenth 90, 146
    to nearest whole number 90, 146
    to one decimal place 90
    to two decimal places 90
    up 90

sample size 198, 200
sample space diagram 30
scale factor 114
scales 64
secondary source 198
sequences
    ascending 10
    consecutive terms 10, 12, 156
    descending 10
    general term 152, 154, 156
    generating 8, 10, 152, 158
    investigating 12
    nth term 152, 154, 156
    patterns 8, 154
    spreadsheets 158
    term-to-term rule 10, 12, 152
    terms 8
shapes
    construction with LOGO 188
    coordinate grid 183
slant height 70
square numbers 9, 140
square root 140
square (shape) 140, 183

square-based pyramid 190, 194
statistics 134, 138, 210
steeper gradient 84
straight-line graphs 78, 80, 82, 84
strategies for problem solving 174, 176, 178
subdivisions on scales 64
substitution
    algebraic expressions 54
    formulae 60
subtraction
    decimals 92
    fractions 44
    integers 4
    inverse 58
    standard written method 92
surface area 197
    cube 72
    cuboid 72
survey 198, 210
symmetry
    line of 108
    mirror line 104, 108
    order of rotation 108
    polygons 108
    reflection 104, 108
    rotation 108
    Wingdings characters 109

tally marks 32
tangram 23
temperature 161
term-to-term rule 10, 12, 152
terms
    algebraic expressions 50
    sequences 8
tessellations 112, 113, 183
tests for divisibility 6
tetrahedron 190, 194
theoretical probability 28, 36
3-D shapes 190, 190
time 64
transformation 110, 112, 114
translation 110
triangles
    angle sum 14
    angles 14, 20, 24
    area 70
    base 70
    drawing 20, 180, 182
    equilateral 24
    exterior angles 17
    height 70
    included angle 180, 182
    included side 180, 182

isosceles 24
protractor 180
right-angled 24
sides 20
triangular numbers 9
triangular prism 190, 194
two-way tables 128

unicursal line 189

vertex 183, 190
vertical axis 204
vertically opposite angles 22, 24
volume
cube 196
cuboid 196

whole numbers 2, 4
Wingdings characters 109

working backwards strategy 176, 178

x-axis 78
x-coordinate 78

y-axis 78, 82
y-coordinate 78
y-intercept 82, 164

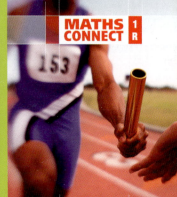

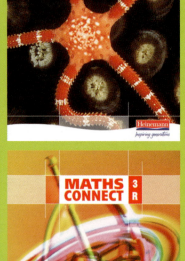
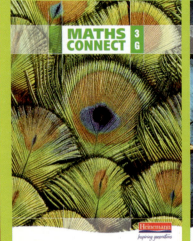
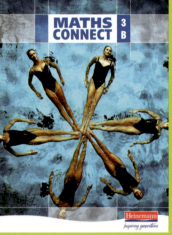
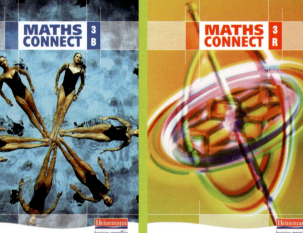
 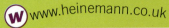